# AutoCAD MEP 2014
# for Designers

**CADCIM Technologies**
*525 St. Andrews Drive*
*Schererville, IN 46375, USA*
*(www.cadcim.com)*

*Contributing Author*
**Sham Tickoo**
*Professor*
*Department of Mechanical Engineering Technology*
*Purdue University Calumet*
*Hammond, Indiana, USA*

# CADCIM Technologies

### AutoCAD MEP 2014 for Designers
### Sham Tickoo

ISBN 978-1-936646-57-9

NOTICE TO THE READER

www.cadcim.com

## DEDICATION

*To teachers, who make it possible to disseminate knowledge
to enlighten the young and curious minds
of our future generations*

*To students, who are dedicated to learning new technologies
and making the world a better place to live in*

## THANKS

*To the faculty and students of the MET department of
Purdue University Calumet for their cooperation*

*To employees of CADCIM Technologies for their valuable help*

# Online Training Program Offered by CADCIM Technologies

CADCIM Technologies provides effective and affordable virtual online training on various software packages including Computer Aided Design and Manufacturing (CAD/CAM), computer programming languages, animation, architecture, and GIS. The training is delivered 'live' via Internet at any time, any place, and at any pace to individuals as well as students of colleges, universities, and CAD/CAM training centers. The main features of this program are:

## Training for Students and Companies in a Classroom Setting

Highly experienced instructors and qualified Engineers at CADCIM Technologies conduct the classes under the guidance of Prof. Sham Tickoo of Purdue University Calumet, USA. This team has authored several textbooks that are rated "one of the best" in their categories and are used in various colleges, universities, and training centers in North America, Europe, and in other parts of the world.

## Training for Individuals

CADCIM Technologies with its cost effective and time saving initiative strives to deliver the training in the comfort of your home or work place, thereby relieving you from the hassles of traveling to training centers.

## Training Offered on Software Packages

CADCIM Technologies provides basic and advanced training on the following software packages:

*CAD/CAM/CAE*: CATIA, Pro/ENGINEER Wildfire/Creo Parametric, SolidWorks, Autodesk Inventor, Solid Edge, NX, AutoCAD, AutoCAD LT, Customizing AutoCAD, AutoCAD MEP, EdgeCAM, and ANSYS

*Computer Programming*: C++, VB.NET, Excel, Dreamweaver, Oracle, AJAX, and Java

*Animation and Styling*: Autodesk 3ds Max, 3ds Max Design, Maya, eyeon Fusion, The Foundry Nuke, and Alias Design

*Architecture and GIS*: Autodesk Revit Architecture, Revit Structure, Navisworks, Raster Design, AutoCAD Civil 3D, and AutoCAD Map 3D

For more information, please visit the following link:

*http://www.cadcim.com*

## Note

If you are a faculty member, you can register by clicking on the following link to access the teaching resources: *http://www.cadcim.com/Registration.aspx*. The student resources are available at *http://www.cadcim.com*. We also provide **Live Virtual Online Training** on various software packages. For more information, write us at *sales@cadcim.com*.

# Table of Contents

## Chapter 4: Creating the HVAC System

## Chapter 5: Creating Piping System

## Chapter 6: Creating Plumbing System

# Chapter 7: Creating Electrical System Layout

# Chapter 8: Representation and Schedules

## Chapter 9: Working with Schematics

## Chapter 10: Project

**Index**

# **Preface**

## AutoCAD MEP 2014

AutoCAD MEP, also known as AMEP, is an AutoCAD Architecture-based software. Here MEP stands for Mechanical Electrical and Plumbing. Therefore, the software has all the required features for creating a Mechanical, Electrical, and Plumbing system. All these systems together make a project. As the software is developed using AutoCAD-based platform, it includes all the features of the base platform such as Blocks, Layers, 3D Models, and so on. It also includes Architectural features such as walls, doors, windows, and so on.

In addition to creating various systems in AMEP, you can add objects with actual parameters in the project. These objects are available in various categories of AutoCAD MEP library and can be customized according to the user requirement.

**AutoCAD MEP 2014 for Designers** textbook is written with the intention of helping the readers effectively use the AutoCAD MEP 2014 designing and drafting tools. This textbook provides a simple and clear explanation of tools that are commonly used in AutoCAD MEP 2014. After reading this textbook, you will be able to design HVAC system, piping system, plumbing system, and electrical layout of a building. The chapter on schematics will enable the users generate the schematic drawings of a system for easy representation. The examples and tutorials used in this textbook ensure that the users can relate the knowledge of this textbook with the actual industry designs.

The main features of this textbook are as follows:

- **Tutorial Approach**
  The author has adopted the tutorial point-of-view and the learn-by-doing approach throughout the textbook. This approach guides the users easily understand the process of designing and drafting with the help of tutorials.

- **Real-World Projects as Tutorials**
  The author has used about 28 real-world mechanical engineering projects as tutorials in this textbook. This enables the readers to relate the tutorials to the engineering industry. In addition, there are about 20 exercises that are also based on the real-world engineering projects.

- **Tips and Notes**
  Additional information related to various topics is provided to the users in the form of tips and notes.

- **Heavily Illustrated Text**
  The text in this book is heavily illustrated with about 400 line diagrams and screen capture images.

- **Learning Objectives**
  The first page of every chapter summarizes the topics that are covered in that chapter.

- **Self-Evaluation Test, Review Questions, and Exercises**
  Every chapter ends with a Self-Evaluation test so that the users can assess their knowledge of the chapter. The answers to the Self-Evaluation test are given at the end of the chapter. Also, the Review Questions and Exercises are given at the end of each chapter and they can be used by the Instructors as test questions and exercises.

## Symbols Used in this Textbook

**Note**

The author has provided additional information to the users about the topic being discussed in the form of Notes.

**Tip**

Special information and techniques are provided in the form of Tips that will increase the efficiency of the users.

## Formatting Conventions Used in the Textbook

Please refer to the following list for the formatting conventions used in this textbook.

- Names of tools, buttons, options, and palettes are written in boldface.

  Example: The **Wall** tool, the **OK** button, the **Left** option, and so on.

- Names of dialog boxes, drop-downs, drop-down lists, list boxes, areas, edit boxes, check boxes, radio buttons, and palettes are written in boldface.

  Example: The **Detail Component Manager** dialog box, the **Walls** drop-down, the **Width** edit box in the **Properties** palette, the **Schematic Curve** check box in the **Drafting Settings** dialog box, the **Keyboard entry** radio button of the **User Preferences** tab in the **Options** dialog box, and so on.

- Values entered in edit boxes are written in boldface.

  Example: Enter **5** in the **Radius** edit box.

- Names and paths of the files are written in italics.

  Example: *C:\amep\c03*, *c03tut03.dwg*, and so on

# Naming Conventions Used in the Textbook
## Tool
If you click on an item in a **Ribbon** and a command is invoked to create/edit an object or perform some action, then that item is termed as **tool**.

For example:
To Create: **Line** tool, **Dimension** tool, **Wall** tool
To Modify: **Move** tool, **Explode** tool, **Rotate** tool
Action: **Zoom All** tool, **Pan** tool, **Copy** tool

If you click on an item in the **Ribbon** and a dialog box is invoked wherein you can set the properties to create/edit an object, then that item is also termed as **tool**, refer to Figure 1.

For example:
**Air Handler** tool, **Pump** tool, **Junction Box** tool

## Button
The item in a dialog box that has a 3D shape like a button is termed as **Button**. For example, **OK** button, **Cancel** button, **Apply** button, and so on.

## Dialog Box
The naming conventions used for the components in a dialog box are mentioned in Figure 1.

***Figure 1*** *The components of a dialog box*

## Tool Palette

A tool palette is the one in which a set of common tools are grouped together for performing an action. For example, **Tool Palettes - Piping**, **Tool Palettes - Architectural**, **Tool Palettes - HVAC**, and so on; refer to Figure 2.

## Properties Palette

The **Properties** palette looks similar to the tool palette but in this palette, only the properties of the objects are displayed. You can edit these properties as per your requirement. Figure 3 shows the **Properties** palette displayed after selecting a wall from the drawing area.

*Figure 2 Tools in the **Tool Palettes - Architectural***    *Figure 3 Properties in the **Properties** palette*

## Free Companion Website

It has been our constant endeavor to provide you the best textbooks and services at affordable price. In this endeavor, we have come out with a Free Companion website that will facilitate the process of teaching and learning of AutoCAD MEP. If you purchase this book, you will get access to the files on the Companion website by visiting *www.cadcim.com*.

The following resources are available for the faculty and students in this website:

### Faculty Resources

- **Technical Support**
  You can get online technical support by contacting ***techsupport@cadcim.com***.

- **Instructor Guide**
  Solutions to all review questions and exercises in the textbook are provided in this guide to help the faculty members test the skills of the students.

- **PowerPoint Presentations**
  The contents of the book are arranged in PowerPoint slides that can be used by the faculty for their lectures.

- **Part Files**
  The part files used in illustrations, tutorials, and exercises are available for free download.

  Note that you can access the faculty resources only if you are registered as faculty at *http://cadcim.com/Registration.aspx*

## Student Resources

- **Technical Support**
  You can get online technical support by contacting ***techsupport@cadcim.com***.

- **Part Files**
  The part files used in illustrations and tutorials are available for free download.

## Stay Connected

To get the latest information about our textbooks, videos, teaching/learning resources, and other such updates, follow us on Facebook (***www.facebook.com/cadcim***) and Twitter (***@cadcimtech***). For more information about our latest video tutorials, you can also subscribe to our YouTube channel (***www.youtube.com/cadcimtech***).

You can access additional learning resources by visiting ***http://allaboutcadcam.blogspot.com***.

If you face any problem in accessing these files, please contact the publisher at ***sales@cadcim.com*** or the author at ***stickoo@purduecal.edu*** or ***tickoo525@gmail.com***.

# Chapter 1

# Introduction to AutoCAD MEP

## Learning Objectives

**After completing this chapter, you will be able to:**

- *Start AutoCAD MEP*
- *Use the components of the initial AutoCAD MEP screen*
- *Invoke AutoCAD MEP commands from the keyboard, menu, toolbar, shortcut menu, Tool Palettes, and Ribbon*
- *Use the components of dialog boxes in AutoCAD MEP*
- *Start a new drawing*
- *Save a work using various file-saving commands*
- *Close a drawing*
- *Open an existing drawing*
- *Exit AutoCAD MEP*
- *Use various options of AutoCAD MEP help*
- *Use Active Assistance, Learning Assistance, and other interactive help topics*

# INTRODUCTION

AutoCAD MEP is a software built on the base of AutoCAD Architecture. Since it belongs to AutoCAD family, it has all the features of AutoCAD like Blocks, Layers, 3D Models, and so on. Also, the software has all the important features of Architecture like, Walls, Doors, Windows, and so on. AutoCAD MEP is also referred as AMEP where MEP stands for Mechanical, Electrical and Plumbing. Therefore, the software has all the required features for creating a Mechanical, Electrical, and Plumbing system. In this chapter, you will learn to start AutoCAD MEP and use various components displayed in AutoCAD MEP interface.

# STARTING AutoCAD MEP

After you have installed AutoCAD MEP 2014, three icons of AutoCAD MEP 2014 are displayed on the desktop named as AutoCAD MEP 2014 - English (Global), AutoCAD MEP 2014 - English (US Imperial), and AutoCAD MEP 2014 - English (US Metric). You can start AutoCAD MEP by double-clicking on any of the three icons. If you double click on AutoCAD MEP 2014 -English (US Imperial) then the units available in drawing will be in Inch, Feet, and Mile. If you double click on AutoCAD MEP 2014 - English (US Metric) then the units available in the drawing will be in Millimeter, Centimeter, and Meter. If you double click on AutoCAD MEP 2014 - English (Global) then AutoCAD MEP starts using the global template. You can also load AutoCAD MEP from the Windows taskbar. To do so, choose **Start > All Programs > Autodesk > AutoCAD MEP 2014 - English** folder to display AutoCAD MEP programs and then choose **AutoCAD MEP 2014 - English** to start AutoCAD MEP, refer to Figure 1-1. On starting AutoCAD MEP, the initial screen will be appeared along the Welcome window, refer to Figure 1-2.

***Figure 1-1*** *Starting AutoCAD MEP 2014 from the Windows taskbar*

**Note**

*In this textbook, the global unit system is followed, so you need to start AutoCAD MEP 2014 by double-click on the AutoCAD MEP 2014 - English (Global) icon from the desktop.*

In the **Welcome** window, you can access information on enhancements, new features, products and services, subscription, are available. You can click on a link to find information available in that link. For example, in the **What's New in 2014** area, you can learn about new features and enhancements of AutoCAD MEP 2014. Choose the **Close** button from this window to exit, the AutoCAD MEP screen will be activated.

*Figure 1-2  The **Welcome** window of AutoCAD MEP 2014*

# AutoCAD MEP SCREEN COMPONENTS

Various components of the initial AutoCAD MEP screen are drawing area, command window, menu bar, toolbars, model and layout tabs, and status bar, refer to Figure 1-3. A title bar that has AutoCAD MEP symbol and the current drawing name is displayed on top of the screen.

## Drawing Area

The drawing area covers a major portion of the screen. In this area, you can draw the objects and use the commands. To draw the objects, you need to define the coordinate points. Position of the pointing device is represented on the screen by the cursor. There is a coordinate system icon at the lower left corner of the drawing area. The drawing area also has the standard windows buttons such as close, minimize, and scroll bar on the top right corner. These buttons have the same functions as for any other standard window.

## Command Window

The command window at the bottom of the drawing area has the Command prompt where you can enter the commands. It also displays the subsequent prompt sequences and the messages.

You can change the size of the window by placing the cursor on the top edge (double line bar known as the grab bar) and then dragging it. This way you can increase its size to see all the previous commands you have used. By default, the command window displays only three lines. You can also press the F2 key to display **AutoCAD Text window**, which displays the previous commands and prompts.

*Figure 1-3* *AutoCAD MEP screen components*

**Tip**
*You can increase the display area of the screen by pressing the CTRL+0 keys or by choosing* ***View > Clean Screen*** *from the menu bar. Note that the **0** key on the numeric keypad of the keyboard cannot be used for the **Clean Screen** option. You can also choose the **Clean Screen** button in the **Application Status Bar** to hide all toolbars.*

## ViewCube
ViewCube is available on the top right corner of the drawing area and is used to switch between the standard and isometric views or roll the current view.

## In-Canvas Viewport Controls
**In-Canvas Viewport Controls** is available on the top left corner of the drawing screen. It enables you to change the view, visual style as well as the viewport.

## Status Bar
The Status Bar displayed at the bottom of the screen is called **Application Status Bar**. It contains some useful information and buttons, refer to Figure 1-4, that help to change the status of some AutoCAD MEP functions. You can toggle between the on and off states of most of these functions by using the corresponding options. Some of the options in the **Application Status Bar** are discussed next.

*Figure 1-4* *The **Application Status Bar***

## Drawing Coordinates

The information about the coordinates is displayed on the left-corner of the Status Bar. The **COORDS** system variable controls the type of display of the coordinates. If the value of the **COORDS** variable is set to 0, the coordinate display is static, that is, the coordinate values displayed in the Status Bar will change only when you specify a point. If the value of the **COORDS** variable is set to 1 or 2, the coordinate display is dynamic. When the variable is set to 1, AutoCAD MEP constantly displays the absolute coordinates of the graphics cursor with respect to the UCS origin. The polar coordinates (length<angle) are displayed when you are in an AutoCAD MEP command and the **COORDS** variable is set to 2. Click on the **Drawing Coordinates** area to toggle the coordinate status from on to off and vice versa.

## Infer Constraints

If this button is chosen, then some of the geometric constraints will be automatically applied to the sketch while it is drawn.

## Snap Mode

On choosing this button, you can move the cursor in fixed increments. The F9 key acts as a toggle key to turn the snap off or on.

## Grid Display

The **Grid Display** button is used to toggle the display of grid lines on and off in the screen. In AutoCAD MEP, the grid lines are used as reference lines to draw objects. The F7 function key can be used to turn the grid display on or off.

## Ortho Mode

On choosing the **Ortho Mode** button, you can draw lines at right angles only. You can use the F8 function key to turn the ortho mode on or off.

## Polar Tracking

The **Polar Tracking** button is used to turn the polar tracking on. If you turn the polar tracking on, the movement of the cursor is restricted along a path based on the angle set as the polar angle. You can also use the F10 function key to turn on this option. Note that turning the polar tracking on, automatically turns off the ortho mode.

## Object Snap

On choosing the **Object Snap** button, you can use the running object snaps to snap on to a point. You can also use the F3 function key to turn the object snap on or off. The status of **OSNAP** (off or on) does not prevent you from using the immediate mode object snaps.

## 3D Object Snap

On choosing this button, you can snap the key point on a solid or a surface. You can also use the F4 function key to turn on or off the 3D object snap.

## Object Snap Tracking

This button is used to turn the object snap tracking on or off. On choosing this button, the inferencing lines will be displayed. Inferencing lines are dashed lines that are displayed automatically when you select a sketching tool and track a particular key point on the screen. You can also choose the F11 function key to turn on or off the object snap tracking.

## Allow/Disallow Dynamic UCS

Choosing this button enable or disable the use of dynamic UCS. Allowing the dynamic UCS ensures that the XY plane of the UCS got dynamically aligned with the selected face of the model. You can also use the F6 function key to turn the **Dynamic UCS** button on or off.

## Dynamic Input

The **Dynamic Input** button is used to turn the **Dynamic Input** mode on or off. Turning it on facilitates the heads-up design approach because all commands, prompts, and dimensional inputs will now be displayed in the drawing area and you do not need to look at the Command prompt all the time. This saves the design time and also increases the efficiency of the user. If the **Dynamic Input** mode is turned on, you will be able to enter the commands through the **Pointer Input** boxes, and the numerical values through the **Dimensional Input** boxes. You will also be able to select the command options through the **Dynamic Prompt** options in the graphics window. To turn the **Dynamic Input** mode on or off, use the F12 key.

## Show/Hide Lineweight

This button is used to turn on or off the display of line weights in the drawing. If this button is not chosen, the display of lineweight will be turned off.

## Show/Hide Transparency

This button is used to turn on or off the transparency set for a drawing. You can set the transparency in the **Properties** panel or in the layer in which the sketch is drawn.

## Quick Properties

On choosing this button, the properties of the selected sketched entity will be displayed in a panel.

## Selection Cycling

On choosing this button, you can cycle through and select the overlapping objects close to the other entities. On selecting an entity when this button is chosen, the **Selection** list box will be displayed with a list of entities.

## Annotation Monitor

The **Annotation Monitor** button is used to turn the **Annotation Monitor** on or off. If it is turned on, all the non-associative annotations will get highlighted with a badge placed on them, as shown in Figure 1-5. In this figure, a line leader is not associated with line.

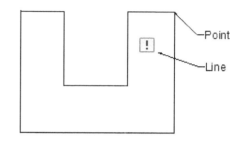

*Figure 1-5* *The non-associative annotation*

## Model or Paper space

The **Model & Paper space** button is used to toggle between the model space and the paper space

## Model

The **Model** button is chosen by default. As a result, you can directly work in the model space to create the drawings.

## Work

The **Work** button can be chosen if you need to work in the paper space. In paper space, you can create viewports, notes, dimensions and so on.

## Quick View Layouts

This button is used to display a panel from which you can choose the desired layout.

## Quick View Drawings

This button is used to display a panel from which you can choose the desired drawings.

## Workspace

When you choose this button, a flyout is displayed. You can use the options in this flyout to switch between workspaces like HVAC, Piping, Electrical, and so on. You can also customize a workspace or create a new workspace by using the options in this flyout.

## Toolbar/Window Positions Unlocked

The **Toolbar/Window Positions Unlocked** button is used to lock and unlock the positions of toolbars and windows. When you click on this button, a shortcut menu is displayed. If you choose the **Floating Toolbars/Panels** option the current position of the floating toolbars will be locked. A check mark will be displayed in the shortcut menu on the type of toolbars that are currently locked. Choosing the **Docked Toolbars/Panels** option from the shortcut menu allows you to lock the current position of all the docked toolbars. Similarly, you can lock or unlock the position of floating and docked windows, such as the **Properties** window or the **Tool Palettes**. If you move the cursor on the **All** option, a cascading menu is displayed that provides the option to lock and unlock all the toolbars and windows.

**Note**
*The **LOCKUI** system variable is responsible for the locking and unlocking of the toolbars and windows. The following are the values of the system variable:*

*Lockui<0> No toolbar or window locked*
*Lockui<1> Locks all docked toolbars*
*Lockui<2> Locks all docked windows*
*Lockui<4> Locks all floating toolbars*
*Lockui<8> Locks all floating windows*

## Hardware Acceleration On
This button is used to set the performance of the software to an acceptable level.

## Current Elevation
This area displays the value of the current elevation.

## Replace Z value with current elevation
This toggle button is used to replace the elevation value with the specified Z value. When this toggle button is selected, the components will be created on the plane having the elevation equal to the Z value entered for the previously created component.

## Catalog Status
This button is used to check the status of equipment in the MEP database. If the database is not updated then you can regenerate the database by using this button. To do so, double-click on the **Catalog Status** button in the **Application Status Bar**; you will be prompted to specify the name of catalog for which you want to regenerate the database. There are five catalogs in AutoCAD MEP: Cable Tray, Conduit, Duct, Mvpart, and Pipe. You can update any of the catalogs by specifying its name or you can update all the catalogs by specifying **All** at the command prompt.

## Tray Settings
Choose the **Tray Settings** option from the menu displayed on clicking the **Application Status Bar Menu**; the **Tray Settings** dialog box will be displayed. You can control the display of icons and notifications in the tray at the right end of the status bar by selecting appropriate options from this dialog box.

## Clean Screen
This button is used to display the expanded view of the drawing area by hiding all the toolbars except the command window, Status Bar, and menu bar. The expanded view of the drawing area can also be displayed by choosing **View > Clean Screen** from the menu bar or by using the CTRL+0 keys. Choose the **Clean Screen** button again to restore the previous display state.

## Status Toggles

You can hide the display of some of the buttons in the Status Bar. To do so, right-click on the **Application Status Bar**; a shortcut menu will be displayed. Move the cursor on the **Status Toggles** option in the shortcut menu; a cascading menu will be displayed. Clear the check mark near the names of the corresponding buttons in the cascading menu to hide their display.

## Plot/Publish Details Report Available

This icon is displayed at the bottom right corner of the window when a plotting or publishing activity is being performed in the background. When you click on this icon, the **Plot and Publish Details** dialog box, which provides the details about the plotting and publishing activity, will be displayed. You can copy this report to the clipboard by choosing the **Copy to Clipboard** button from the dialog box.

## Manage Xrefs

The **Manage Xrefs** icon is displayed whenever an external reference drawing is attached to the selected drawing. This icon displays a message and an alert whenever the Xreffed drawing needs to be reloaded. To find detailed information regarding the status of each Xref in the drawing and the relation between the various Xrefs, click on the **Manage Xrefs** icon; the **External References Palette** will be displayed.

# Drawing Status Bar

The **Drawing Status Bar** is displayed in between the drawing area and the command window by default, refer to Figure 1-6. Click on the **Application Status Bar Menu** arrow and choose the **Drawing Status Bar** option from the flyout; the **Drawing Status Bar** will hide. Also, the **Drawing Status Bar** and the **Annotation Scale**, **Annotation Visibility**, and **Automatically Add Scale** buttons will move automatically to the **Application Status Bar**. If you turn on the **Drawing Status Bar**, these buttons will move back to the **Drawing Status Bar**. The annotated **Drawing Status Bar** is shown in Figure 1-7. The buttons available in the **Drawing Status Bar** are discussed next.

*Figure 1-6 The Drawing Status Bar*

*Figure 1-7  The annotated **Drawing Status Bar***

## Annotation Scale
When you click on this button, a flyout will be displayed. The options available in this flyout are used to set the scale of the annotative objects inserted in the paper layout.

## Annotation Visibility
This button is used to control the visibility of the annotative objects that do not support the current annotation scale in the drawing area.

## Automatically Add Scale
If this button chosen, then the annotation scales that are set current are added to all the annotative objects present in the drawing automatically.

## Level of Detail
When you click on this button, a flyout will be displayed. Various options available in this flyout are used to control the level of details of the objects created in the drawing area. The options in the flyout are arranged according to their area of application. For example, the options related to the **Electrical** workspace are arranged together.

## Cut Plane
This option is used to specify the cut plane height and display range for objects in the drawing area. If you click on the value displayed next to **Cut Plane** in the **Drawing Status Bar**; the **Global Cut Plane** dialog box will be displayed, as shown in Figure 1-8.

## Surface Hatch Toggle
Using this button, you can turn on/off the display of the surface hatch in the drawing area.

## Layer Key Overrides
The **Layer Key Overrides** button is used to enable or disable the overrides applied on the layers available in the drawing file. When you click on this button, the **Layer Key Overrides** dialog box is displayed, refer to Figure 1-9. Using the options available in this dialog box, you

can configure the presets for any of the layers available in the drawing. Also, you can enable or disable the overrides for any of the layer available.

*Figure 1-8* *The* ***Global Cut Plane*** *dialog box*

*Figure 1-9* *The* ***Layer Key Overrides*** *dialog box*

## AEC Project Standards

Using this button, you can configure the standards for current loaded project. When you click on this button, a flyout will be displayed with two options: **Synchronize Drawing** and **Configure Standards.** On choosing the **Configure Standards** option, the **Configure AEC Project Standards** dialog box will be displayed, refer to Figure 1-10. The **Standard Styles** tab is chosen by default in the dialog box. You can standardize any of the objects available in the **Objects** list of this tab by selecting the check box corresponding to the object in the dialog box. If you choose the **Synchronization** tab, three radio buttons will be displayed: **Automatic**, **Semi-automatic**, and **Manual**. The **Manual** radio button is selected by default. So, all the drawing standards need to be applied manually. Project drawings will not be synchronized with the project standards. If you select the **Semi-automatic** radio button, the project standards will be applied only when the project drawing is opened. Also, you will be prompted to apply project standard on each opened project drawing. If you select the **Automatic** radio button, the project standards will be applied on each opened drawing without any prompt. Using the options in the **AutoCAD Standards** tab, you can load any AutoCAD Standard file to apply as standard on the project drawings.

*Figure 1-10  The **Configure AEC Project Standards** dialog box*

## Isolate Objects

When you click on this button, a flyout is displayed. There are two buttons in this flyout: **Isolate Objects** and **Hide Objects**. Using the **Isolate Objects** button, you can isolate the selected objects so that only the selected objects will be displayed in the drawing area. If there are some objects already isolated, then the **Isolate Objects** button will be displayed as a red bulb in the **Drawing Status Bar**. To end isolation or display a hidden object, click this button again and choose the **End Object Isolation** option. To hide objects, choose the **Hide Objects** option from the flyout.

## Drawing Status Bar Menu

This button is available in the right of the **Drawing Status Bar**. When you click on this button, the **Drawing Status Bar** menu will be displayed, as shown in Figure 1-11. You can hide or display any of the button in the **Drawing Status Bar** by using the respective check box in the menu.

*Figure 1-11  The **Drawing Status Bar** menu*

# INVOKING COMMANDS IN AutoCAD MEP

On starting AutoCAD MEP, when you are in the drawing area, you need to invoke AutoCAD MEP commands to perform any operation. For example, to draw a pipe line, enter the **PIPEADD** command at the command prompt and then define the start point and the endpoint of the pipe. Similarly, if you want to erase objects, you must invoke the **ERASE** command and then select the objects for erasing. In AutoCAD MEP, you can invoke the commands by using:

| | | | |
|---|---|---|---|
| **Command Line** | **Ribbon** | **Application menu** | **Tool Palettes** |
| **Tool Palettes** | **Menu bar** | **Shortcut menu** | **Toolbar** |

## Command Line

You can invoke any AutoCAD MEP command from the Command line by typing the command name and then pressing ENTER. As you type the first letter of command, AutoCAD MEP displays all available commands starting with the letter typed. If the **Dynamic Input** is on and the cursor is in the drawing area, by default, the command will be entered through the **Pointer Input** box. The **Pointer Input** box is a small box displayed on the right of the cursor. However, if the cursor is currently placed on any toolbar or menu bar, or if the **Dynamic Input** is turned off, the command will be entered through the Command prompt. Before you enter a command, the Command prompt is displayed as the last line in the command window area. If it is not displayed, you must cancel the existing command by pressing the ESC (Escape) key. The following example shows how to invoke the **LINE** command using the keyboard:

  Command: **LINE** or **L** [Enter]   (L is command alias)

## Ribbon

In AutoCAD MEP, you can also invoke a tool from the **Ribbon**. The tools for creating, modifying, and annotating the objects are available in the **Annotate** tab in panels instead of being spread out in the entire drawing area in different toolbars and menus, refer to Figure 1-12.

*Figure 1-12  The **Ribbon** with the **Annotate** tab chosen*

When you start the AutoCAD MEP session for the first time, by default the **Ribbon** is displayed below the **Quick Access Toolbar**. The **Ribbon** consists of various tabs. The tabs have different panels, which in turn, have tools arranged in rows. Some of the tools have small black down arrows. This indicates that the tools having similar functions are grouped together. To choose a tool, click on the down arrow; a drop-down will be displayed. Choose the required tool from the drop-down displayed. Note that if you choose a tool from the drop-down, the corresponding command will be invoked and the tool that you have chosen will be displayed in the panel. For example, to draw a circle using 2 points, click on the down arrow next to the **Circle, Center, Radius** tool in the **Draw** panel of the **Home** tab; a flyout will be displayed. Choose the **Circle, 2-Point** tool from the flyout and then draw the circle. You will notice that the **Circle, 2-Point** tool is displayed in place of the **Circle, Center, Radius** tool.

Choose the down arrow to expand the panel. You will notice that a push pin is available at the left corner of the panel. Click on the push pin to keep the panel in the expanded state. Also, some of the panels have an inclined arrow at the lower-right corner. When you left click on an inclined arrow, a dialog box is displayed. You can define the setting of the corresponding panel in the dialog box.

You can reorder the panels in the tab. To do so, press and hold the left mouse button on the panel to be moved and drag it to the required position. You can also undock the **Ribbon**. To do so, right-click on the blank space in the **Ribbon** and choose the **Undock** option; the Ribbon gets undocked. You can move, resize, anchor, and auto-hide the **Ribbon** using the shortcut menu that will be displayed when you right-click on the heading strip. To anchor the floating **Ribbon** to the left or right of the drawing area in the vertical position, right-click on the heading strip of the floating **Ribbon**; a shortcut menu is displayed. Choose the corresponding option from this shortcut menu. The **Auto-hide** option will hide the **Ribbon** into the heading strip and will display it only when you move the cursor over this strip.

You can customize the display of tabs and panels in the **Ribbon**. To customize the **Ribbon**, right-click on any one of the tools in it; a shortcut menu will be displayed. On moving the cursor over one of the options, a flyout will be displayed with a tick mark before all options and the corresponding tab or panel will be displayed in the **Ribbon**. Select/clear appropriate option to display/hide a particular tab or panel.

## Application Menu

The **Application** menu is available at the top-left of the AutoCAD MEP window. It contains some of the tools that are also available in the **Quick Access Toolbar**. Click the down arrow on the **Application** menu to display the tools, as shown in Figure 1-13. You can search a command using the search field on the top of the **Application** menu. To search a tool, enter the complete or partial name of the command in the search field; the relevant tool list will be displayed. If you click on a tool from the list, the corresponding command will get activated.

By default, the **Recent Document** button is chosen in the **Application** menu. Therefore, the recently opened drawings will be displayed. If you have opened multiple drawing files, choose the **Open Documents** button; the documents that are opened will be listed in the **Application** menu. To set the preferences of the file, choose the **Options** button available at the bottom-right of the **Application** menu. To exit AutoCAD MEP, choose the **Exit AutoCAD MEP 2014** button next to the **Options** button.

*Figure 1-13  The **Application** menu*

## Tool Palettes

AutoCAD MEP has provided **Tool Palettes** as an easy and convenient way of placing and sharing hatch patterns and blocks in the current drawing. By default, the **Tool Palettes** are displayed on the left in the drawing area. If the **Tool Palettes** are not displayed then choose **Tools** from the **Tools** drop-down in the **Build** panel of the **Home** tab in the **Ribbon** or choose the CTRL+3 keys to display the **Tool Palettes**. You can resize the **Tool Palettes** using the resizing cursor that is displayed when you place the cursor on the top or bottom extremity of the **Tool Palettes**. The **Tool Palettes** will be discussed in detail in later chapters.

## Menu Bar

You can also select commands from the menu bar. Menu bar is not displayed by default. To display the menu bar, choose the down arrow in the **Quick Access Toolbar**; a flyout is displayed.

Choose the **Show Menu Bar** option from it; the menu bar will be displayed. As you move the cursor over the menu bar, different titles are highlighted. You can choose the desired item by clicking on it; the corresponding menu is displayed directly under the title. You can invoke a command by left-clicking on a menu. In AutoCAD MEP, there are only three menus available: **File**, **Window**, and **Help**. The **File** menu has the options to manage the drawing file. Using the options available in the **Window** menu, you can close the current session, switch between two sessions or can display multiple sessions in the display window altogether. The options in the **Help** menu are used to display the help documentation.

## Shortcut Menu

AutoCAD MEP has provided shortcut menus as an easy and convenient way of invoking the recently used tools. These shortcut menus are context-sensitive, which means that the tools present in them are dependent on the place/object for which they are displayed. A shortcut menu is invoked by right-clicking and is displayed at the cursor location. You can right-click anywhere in the drawing area to display the general shortcut menu. It generally contains an option to select the previously invoked tool again, apart from the common tools, refer to Figure 1-14.

***Figure 1-14*** *Shortcut menu with the recently used commands*

If you right-click in the drawing area while a command is active, a shortcut menu, with the options related to that particular command, is displayed. Figure 1-15 shows the shortcut menu when the **Duct** tool is active.

If you right-click on the **Layout** tab, a shortcut menu will be displayed that contains the options for layouts, refer to Figure 1-16.

You can also right-click on the command window to display the shortcut menu. This menu displays the six most recently used commands and some of the window options like **Copy** and **Paste**, refer to Figure 1-17. The commands and their prompt entries are displayed in the **History** window (previous command lines not visible) and can be selected, copied, and pasted in the command line using the shortcut menu. As you press the up arrow key, the previously entered commands are displayed in the command window. Once the desired command is displayed at the Command prompt, you can execute it by simply pressing the ENTER key. You can also copy and edit any previously invoked command by locating it in the **History** window and then selecting the command lines. After selecting the desired command lines from the

**History** window, right-click to display a shortcut menu. Choose **Copy** from the menu and then paste the selected lines at the end of the command line.

You can right-click on the coordinate display area of the **Application Status Bar** to display the shortcut menu. This menu contains the options to modify the display of coordinates, as shown in Figure 1-18.

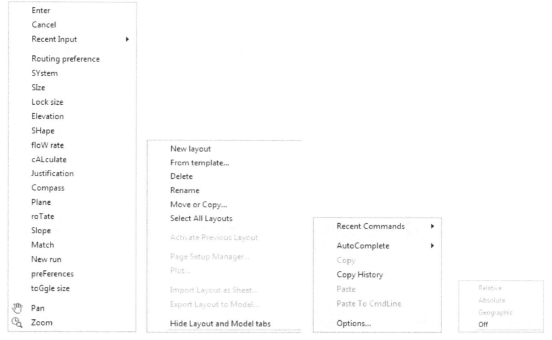

*Figure 1-15* *Shortcut menu when the* **DUCT** *tool is active*

*Figure 1-16* *Shortcut menu for the* **Layout** *tab*

*Figure 1-17* *Command line window shortcut menu*

*Figure 1-18* *The Status Bar shortcut menu*

# AutoCAD MEP DIALOG BOXES

There are certain commands, which when invoked, display a dialog box. A dialog box is one of the convenient methods of user interface. When you choose an item in a menu with the ellipses [...], it displays the dialog box. For example, when you choose **New** from the **File** menu, the **Select Template** dialog box is displayed. In the **Application** menu, when you choose an item it may display a dialog box although there are no ellipses with that item. For example, when you choose **Options** from the **Application** menu, the **Options** dialog box is displayed. A dialog box contains a number of parts like the dialog label, radio buttons, text or edit boxes, check boxes, slider bars, image boxes, and command buttons. These components are also referred to as tiles. Some of the components of a dialog box are shown in Figure 1-19.

You can select the desired tile using the pointing device, which is represented by an arrow when a dialog box is invoked. The titlebar displays the name of the dialog box. The tabs specify the various sections with a group of related options under them. The check boxes are toggle options for making a particular option available or unavailable. The drop-down list displays an item and an arrow on the right which when selected displays a list of items to choose from. You can make a selection in the radio buttons. Only one can be selected at a time. The preview displays the preview image of the item selected. The text box is an area where you can enter a text like a file name. It is also called an edit box, because you can make any change to the text entered. In some dialog boxes, there is the **[...]** button, which displays another related dialog box. There are certain buttons (**OK**, **Cancel**, **Help**) at the bottom of the dialog box. The name implies their functions. The button which is highlighted is the default button. The **Help** button in this dialog box is used to display help on the various features of the dialog box.

*Figure 1-19  Components of a dialog box*

# STARTING A NEW DRAWING

| | | | |
|---|---|---|---|
| **Application Menu:** | New > Drawing | **Command:** | NEW |
| **Quick Access Toolbar:** | New | **Menu Bar:** | File > New |

You can open a new drawing using the **New** tool in the **File** menu. When you invoke the **New** tool, by default the **Select template** dialog box will be displayed, as shown in Figure 1-20. This dialog box displays a list of default templates available in AutoCAD MEP 2014. The default selected template is *Aecb Model (Global Ctb).dwt*, which starts the AutoCAD MEP environment with global unit system. You can select any other template to start a new drawing. The drawing will use the settings of the selected template. You can also open any drawing without using any template either in metric or imperial system. To do so, choose the down arrow on the right of the **Open** button from the **Select Template** dialog box and select the **Open with no Template-Metric** option or the **Open with no Template-Imperial** option from the flyout.

You can also open a new drawing using the **Use a Wizard** and **Start from Scratch** options from the **Create New Drawing** dialog box. To invoke the **Create New Drawing** dialog box, enter **STARTUP** in the command window and then enter **1** as the new value for this system variable. Invoke the **New** tool; the **Create New Drawing** dialog box will be displayed, as shown in Figure 1-21. The options in this dialog box are discussed next.

*Figure 1-20* *The* *Select template* *dialog box*

*Figure 1-21* *The* *Create New Drawing* *dialog box*

**Note**
*If you have started a new AutoCAD MEP session with the **STARTUP** variable set to 1, then the **Startup** dialog box is displayed. The options in the **Startup** dialog box is same as that of*

*the **Create New Drawing** dialog box with the only difference that in the **Startup** dialog box, the **Open a Drawing** button will be activated.*

## Open a Drawing
By default, this option is not available. You can access this option only when you start a new session of AutoCAD MEP. You can open a drawing by using the **Open** button which is discussed later in this chapter.

## Start from Scratch
When you choose the **Start from Scratch** button, refer to Figure 1-21, AutoCAD MEP provides you with options to start a new drawing that contains the default AutoCAD MEP setup for Imperial or Metric drawings. If you select the Imperial default setting, the limits are 12X9, text height is 0.20, and dimensions and linetype scale factors are 1.

## Use a Template
When you choose the **Use a Template** button from the **Create New Drawing** dialog box, AutoCAD MEP displays a list of templates, refer to Figure 1-22. The default selected template file is *Aecb model (global ctb).dwt*. You can select any of the template files from the list. The new drawing will have the same settings as specified in the template file. The preview of the template file selected is displayed in the dialog box. You can also define your own template files that are customized to your requirements. To differentiate the template files from the drawing files, the template files have a *.dwt* extension whereas the drawing files have a *.dwg* extension. Any drawing file can be saved as a template file. You can use the **Browse** button to select other template files. When you choose the **Browse** button, the **Select a template file** dialog box is displayed with the **Template** folder open, displaying all the template files.

*Figure 1-22   The default templates displayed in **Create New Drawing** dialog box on choosing the **Use a Template** button*

# Use a Wizard

The **Use a Wizard** button allows you to set the initial drawing settings before actually starting a new drawing. When you choose the **Use a Wizard** button, the **Quick Setup** and **Advanced Setup** options are displayed in the **Select a Wizard** area, refer to Figure 1-23. If you select the **Quick Setup** option, you can specify the units and the limits of the work area. If the **Advanced Setup** option is selected, then you can set the units, limits, and the other types of settings for a drawing. These options are discussed next.

*Figure 1-23* *The wizard options displayed on choosing the* **Use a Wizard** *button*

## Advanced Setup

This option allows you to preselect the parameters of a new drawing such as the units of linear and angular measurements, type and direction of angular measurements, approximate area desired for the drawing, precision for displaying the units after decimal, and so on. When you select the **Advanced Setup** wizard option from the **Create New Drawing** dialog box and choose the **OK** button, the **Advanced Setup** wizard is displayed. The **Units** page is displayed by default, as shown in Figure 1-24.

This page is used to set the units for measurement in the current drawing. The different units of measurement that you can choose from are Decimal, Engineering, Architectural, Fractional, and Scientific. You can select the required unit of measurement by selecting the respective radio button. You will notice that the preview image is modified accordingly. You can also set the precision for the measurement units by selecting it from the **Precision** drop-down list.

Choose the **Next** button to open the **Angle** page, as shown in Figure 1-25. You will notice that an arrow appears on the left of **Angle** in the **Advanced Setup** wizard. This suggests that this page is current.

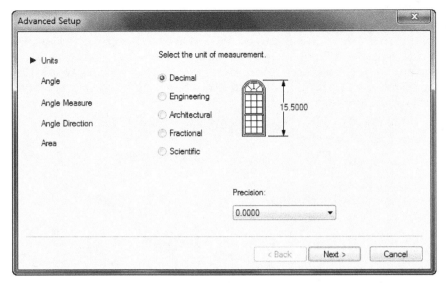

*Figure 1-24  The **Units** page of the **Advanced Setup** wizard*

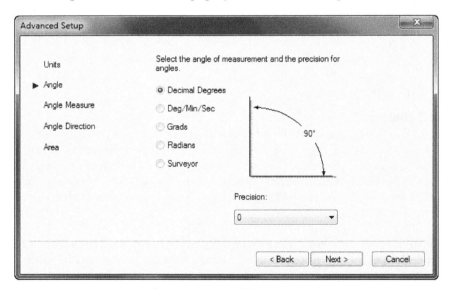

*Figure 1-25  The **Angle** page of the **Advanced Setup** wizard*

This page is used to set the unit for angular measurements and its precision. The units for angle measurement are Decimal Degrees, Deg/Min/Sec, Grads, Radians, and Surveyor. The units for angle measurement can be set by selecting any one of these radio buttons as required. The preview of the selected angular unit is displayed on the right of the radio buttons. The precision format changes automatically in the **Precision** drop-down list depending on the angle measuring system selected. You can then select the precision from the drop-down list.

The next page is the **Angle Measure** page, as shown in Figure 1-26. This page is used to select the direction of the baseline from which the angles will be measured. You can also set your own direction by selecting the **Other** radio button and then entering the value in the edit box displayed below it. This edit box gets activated only when you select the **Other** radio button.

*Figure 1-26  The Angle Measure page of the Advanced Setup wizard*

Choose **Next** to display the **Angle Direction** page to set the orientation for the angle measurement, refer to Figure 1-27. By default the angles are positive, if measured in a counterclockwise direction. This is because the **Counter-Clockwise** radio button is selected. If you select the **Clockwise** radio button, the angles will be considered positive when measured in the clockwise direction.

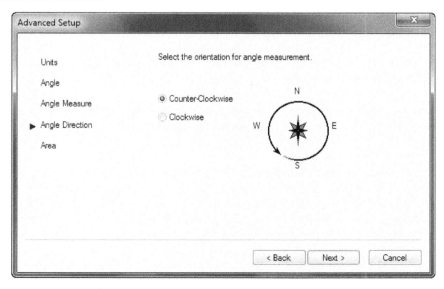

*Figure 1-27  The Angle Direction page of the Advanced Setup wizard*

To set the limits of the drawing, choose the **Next** button; the **Area** page will be displayed, as shown in Figure 1-28. You can enter the width and length of the drawing area in the respective edit boxes.

**Note**
*Even after you increase the limits of the drawing, the drawing display area is not increased. You need to invoke the **Zoom All** tool from the Navigation Bar to increase the drawing display area.*

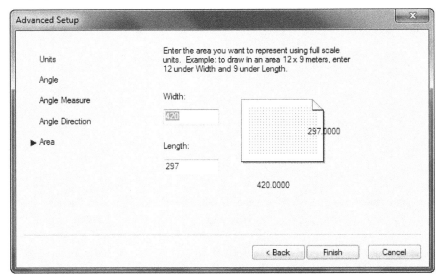

*Figure 1-28  The **Area** page of the **Advanced Setup** wizard*

## Quick Setup

When you select the **Quick Setup** option from the **Create New Drawing** dialog box and choose the **OK** button, the **QuickSetup** wizard is displayed. This wizard has two pages: **Units** and **Area**. The **Units** page is opened by default, as shown in Figure 1-29. The options in the **Units** page are similar to those in the **Units** page of the **Advanced Setup** wizard. The only difference is that you cannot set the precision for the units in this wizard.

Choose **Next** to display the **Area** page, as shown in Figure 1-30. The **Area** page of the **QuickSetup** is similar to that of the **Advanced Setup** wizard. In this page, you can set the drawing limits.

*Tip. When you open an AutoCAD MEP session, a drawing will be opened automatically. But you can open a new drawing using the options such as **Start from Scratch** and **Use a Wizard** from the **Startup** dialog box before entering into AutoCAD MEP environment. As mentioned earlier, the display of the **Startup** dialog box is turned off by default. Refer to the section of Starting a New Drawing to know how to turn on the display of this dialog box.*

*Figure 1-29* *The* **Units** *page of the* **QuickSetup** *wizard*

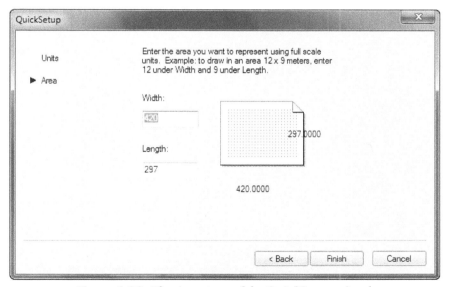

*Figure 1-30* *The* **Area** *page of the* **QuickSetup** *wizard*

# SAVING WORK

You must save your work before you exit the drawing editor or turn off your system. Also, it is recommended that you save your drawings after regular intervals, so that in the event of a power failure or an editing error, all works saved before the problem started could be retained.

AutoCAD MEP has provided the **QSAVE**, **SAVEAS**, and **SAVE** commands that allow you to save your work. These commands allow you to save your drawing by writing it to a permanent storage device, such as a hard drive or any removable drive.

When you choose the **Save** tool from the **Quick Access Toolbar**, the **QSAVE** command is invoked. If you are saving the drawing for the first time in the present session, the **SAVEAS** command will be invoked in place of **QSAVE** command and you will be prompted to enter the file name in the **Save Drawing As** dialog box, as shown in Figure 1-31. You can enter the name for the drawing and then choose the **Save** button. If you have modified a drawing file, choose the **Save** tool to save it; the system saves the file without prompting you to enter a file name. This allows you to do a quick save.

When you choose **Save As** from the **Application** menu or choose the **Save As** tool from the **Quick Access Toolbar**, the **Save Drawing As** dialog box will be displayed, refer to Figure 1-31. Even if the drawing has been saved with a file name, using this tool you can specify a new name for the drawing. You can also use this tool when you make certain changes to a template and want to save the changed template drawing but leave the original template unchanged.

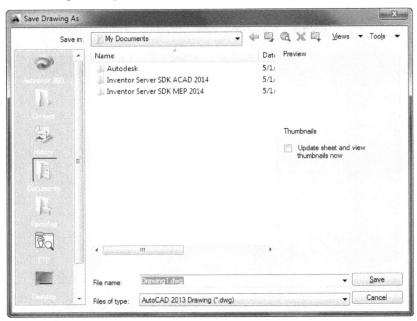

*Figure 1-31  The **Save Drawing As** dialog box*

The options in the **Save Drawing As** dialog box are discussed next.

## Places List
A column of icons is displayed on the left side of the dialog box. These icons contain the shortcuts to the folders that are frequently used. You can quickly save your drawings in one of these folders. The **History** folder displays the list of the most recently saved drawings. You can save your personal drawings in the **Documents** or **Favorites** folder. The **FTP** folder displays

the list of various FTP sites that are available for saving the drawing. By default, no FTP sites are shown in the dialog box. To add an FTP site to the dialog box, choose the **Tools** button on the upper-right corner of the dialog box to display a shortcut menu and select **Add/Modify FTP Locations**. The **Desktop** folder displays the list of contents on the desktop. The **Buzzsaw** icons connect you to their respective pages on the Web. You can add a new folder in this list for an easy access by simply dragging the folder on to the **Places** list area. You can rearrange all these folders by dragging them and then placing them at the desired locations. It is also possible to remove the folders, which are not in frequent use. To do so, right-click on the particular folder and then select **Remove** from the shortcut menu. Now, you can also save the document on Autodesk Cloud. The option for saving the document is discussed next.

The **Autodesk 360** icon is available on the top left in the **Save Drawing As** dialog box, refer to Figure 1-31. It is used to share data online with the users who have an Autodesk account. When you choose this button, the **Autodesk-Sign In** window will be displayed. Now, you can Sign In to upload your document or file in Autodesk Cloud.

### File name Edit Box
To save your work, enter the name of the drawing in the **File name** edit box by typing or by selecting it from the drop-down list. If you have already assigned a name to the drawing, it will be displayed in the edit box as the default name. If the drawing is unnamed, the default name *Drawing1* is displayed in the **File Name** edit box.

### Files of type Drop-down List
The **Files of type** drop-down list is used to specify the drawing format in which you want to save the file, refer to Figure 1-32. For example, to save the file as an AutoCAD 2007 drawing file, select **AutoCAD 2007/LT 2007 Drawing (*.dwg)** from the drop-down list.

*Figure 1-32  The **Files of type** drop-down list*

### Save in Drop-Down List
The current drive and path information is listed in the **Save in** drop-down list. AutoCAD MEP will initially save the drawing in the default folder, but if you want to save the drawing in a different folder, you have to specify the path. For example, to save the present drawing as *house* in the *C1* folder, choose the arrow button in the **Save in** drop-down list to display the drop-down list. Select **C:** from the drop-down list; all folders in the C drive will be listed in the **File** list box. Double-click on the **C1** folder, if it is already listed there or create a new folder C1 by choosing the **Create New Folder** button. Select *house* from the drop-down list, if it is already listed there, or enter it in the **File name** edit box and then choose the **Save** button.

Your drawing (*house*) will be saved in the *C1* folder (*C:\C1\house.dwg*). Similarly, to save the drawing in the D drive, select **D:** from the **Save in** drop-down list.

 **Tip**. *The file name you enter to save a drawing should match its contents. This helps you to remember the drawing details and makes it easier to refer to them later. Also, the file name can be 255 characters long and can contain spaces and punctuation marks.*

## Views Drop-down List
The **Views** drop-down list has the options for the type of listing of files and displaying the preview images, refer to Figure 1-33.

### List, Details, and Thumbnails Options
If you choose the **Details** option, it will display the detailed information about the files (size, type, date, and time of modification) in the **Files** list box. In the detailed information, if you click on the **Name** label, the files are listed with the names in alphabetical order. If you double-click on the **Name** label, the

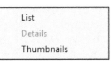

*Figure 1-33  The Views drop-down list*

files will be listed in reverse order. Similarly, if you click on the **Size** label, the files are listed according to their size in ascending order. Double-clicking on the **Size** label will list the files in descending order of size. Similarly, you can click on the **Type** label or the **Modified** label to list the files accordingly. If you choose the **List** option, all files present in the current folder will be listed in the **File** list box. If you select the **Thumbnails** option, the list box displays the preview of all the drawings, along with their names displayed at the bottom of the drawing preview. Also, the preview of the file is displayed in the **Preview** image box.

## Create New Folder
 If you choose the **Create New Folder** button, AutoCAD MEP creates a new folder with the name **New Folder**. The new folder is displayed in the **File** list box. You can change the name of the folder if required.

## Up one level
 When you choose the **Up one level** button, the folders that are up by one level are displayed. For example, if you are in the *Sample* subfolder of the *AutoCAD MEP 2014* folder, then choosing the **Up one level** button will open the *AutoCAD MEP 2014* folder.

## Search the Web
 When you choose this button, the **Browse the Web** dialog box displayed. Using the options in this dialog box, you can access and store AutoCAD MEP files on the Internet. You can also use the ALT+3 keys to browse the Web when this dialog box is available on the screen.

## Tools Drop-down list

The **Add/Modify FTP Locations** option in the **Tools** drop-down list is used for adding or modifying the FTP sites. These sites can then be browsed from the FTP shortcut in the **Places** list. The **Add Current Folder to Places** and **Add to Favorites** options add the folder displayed in the **Save in** edit box to the **Places** list or to the Favorites folder, respectively. The **Options** option displays the **Saveas Options** dialog box where you can save the proxy images of the custom objects. It has the **DWG Options** and **DXF Options** tabs. The **Security Options** option displays the **Security Options** dialog box, which is used to configure the security options of the drawing.

# AUTOMATIC TIMED SAVE

AutoCAD MEP allows you to save your work automatically at specific intervals. To change the time intervals, choose the **Options** button from the **Application Menu**; the **Options** dialog box will be displayed. In this dialog box, enter the duration after which the file will be saved automatically in the **Minutes between saves** text box in the **File Safety Precautions** area of the **Open and Save** tab. This duration depends on the power supply, hardware, and type of drawings. AutoCAD MEP saves the drawing with the file extension .sv$. You can also change the time interval by using the **SAVETIME** system variable.

**Tip**. *Although the automatic save feature saves your drawing after a certain time interval, you should not completely depend on it because the procedure for converting the sv$ file into a drawing file is cumbersome. Therefore, it is recommended that you save your files regularly using the QSAVE or SAVEAS commands.*

# BACKUP FILES

If a drawing file already exists and you use **Save** or **Save As** tools to update the current drawing, AutoCAD MEP creates a backup file. AutoCAD MEP takes the previous copy of the drawing and changes it from .dwg to .bak. The updated drawing is saved as a drawing file with the .dwg extension. For example, if the name of the drawing is *myproj.dwg*, AutoCAD MEP will change it to *myproj.bak* and save the current drawing as *myproj.dwg*.

## Changing Automatic Timed Saved and Backup Files into AutoCAD MEP File Format

In some cases, you may need to change the format of automatic timed saved and backup files into AutoCAD MEP file format. To change the backup file into an AutoCAD MEP file format, open the folder in which you have saved the backup or the automatic timed saved file using **Windows Explorer**. Choose the **Organize > Folder and Search Options** from the menu bar to invoke the **Folder Options** dialog box. Choose the **View** tab and under the **Advanced settings** area, clear the **Hide extensions for known file types** check box, if selected. Exit the dialog box. Rename the automatic saved drawing or the backup file with a different name and also change the extension of the drawing from .sv$ or .bak to .dwg. After you rename the drawing, you will notice that the icon of the automatic saved drawing or the backup file is replaced by the AutoCAD MEP icon. This indicates that the automatic saved drawing or the backup file is changed to an AutoCAD MEP drawing.

# Using the Drawing Recovery Manager to Recover Files

The files that are saved automatically can also be retrieved by using the **Drawing Recovery Manager**. You can open the **Drawing Recovery Manager** by choosing **Drawing Utilities > Open the Drawing Recovery Manager** from the **Application Menu** or by entering **DRAWINGRECOVERY** at the Command prompt.

In case of a system crash, the **Drawing Recovery** message box will be displayed on starting AutoCAD MEP again, refer to Figure 1-34. The message box informs you that the program unexpectedly failed and you can open the most suitable among the backup files created by AutoCAD MEP. Choose the **Close** button from the **Drawing Recovery** message box; the **Drawing Recovery Manager** will be displayed on the left of the drawing area, as shown in Figure 1-35.

**Note**
*The **Drawing Recovery Manager** will be available only when the automatic save feature is active.*

The **Backup Files** rollout lists the original files, the backup files, and the automatically saved files. Select a file; its preview will be displayed in the **Preview** rollout. Also, the information corresponding to the selected file will be displayed in the **Details** rollout. To open a backup file, double-click on its name in the **Backup Files** rollout. Alternatively, right-click on the file name and then choose **Open** from the shortcut menu. It is recommended that you save the backup file at the desired location before you start working on it.

*Figure 1-34  The **Drawing Recovery** message box*

*Figure 1-35  The **Drawing Recovery Manager***

# CLOSING A DRAWING

You can use the **CLOSE** command to close the current drawing file without actually quitting AutoCAD MEP. To do so, choose **Close > Current Drawing** from the **Application Menu** or enter **CLOSE** at the Command prompt; the current drawing file will be closed. If multiple drawing files are opened, choose **Close > All Drawings** from the **Application Menu**. If you have not saved the drawing after making the last changes to it and you invoke the **CLOSE** command, AutoCAD MEP displays a dialog box that allows you to save the drawing before closing. This dialog box gives you an option to discard the current drawing or the changes made to it. It also gives you an option to cancel the command. After closing the drawing, you are still in AutoCAD MEP from where you can open a new or an already saved drawing file. You can also use the **Close** button (**X**) in the drawing area to close the drawing.

**Note**
*You can close a drawing even if a command is active.*

# OPENING AN EXISTING DRAWING

You can open an existing drawing file that has been saved previously. There are three methods that can be used to open a drawing file: by using the **Select File** dialog box, by using the **Create New Drawing** dialog box, and by dragging and dropping.

## Opening an Existing Drawing Using the Select File Dialog Box

If you are already in the drawing editor and you want to open a drawing file, choose the **Open** tool from the **Quick Access Toolbar**; the **Select File** dialog box will be displayed. Alternatively, invoke the **OPEN** command to display the **Select File** dialog box, refer to Figure 1-36. You can select the drawing to be opened using this dialog box. This dialog box is similar to the standard dialog boxes. You can choose the file you want to open from the folder in which it is stored. You can change the folder from the **Look in** drop-down list. You can then select the name of the drawing from the list box or you can enter the name of the drawing file you want to open in the **File name** edit box. After selecting the drawing file, you can choose the **Open** button to open the file.

When you select a file name, preview of the selected file is displayed in the **Preview** box. In this way, you can also use the **Preview** box to find the file of a particular drawing. You can also change the file type by selecting it in the **Files of type** drop-down list. Apart from the *dwg* files, you can open the *dws* (standard), *dwt* (template) files or the *dxf* files. You have all the standard icons in the **Places** list that can be used to open drawing files from different locations. When you click on the down arrow adjacent to the **Open** button, a drop-down list is displayed, as shown in Figure 1-37. You can choose a method for opening the file using this drop-down list. These methods are discussed next.

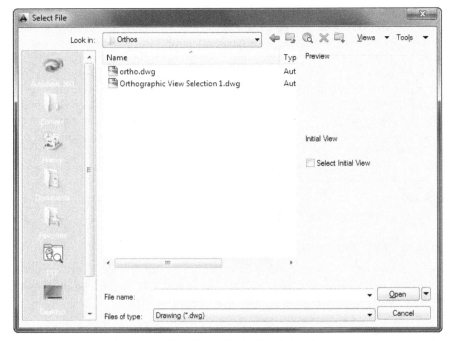

*Figure 1-36*  *The **Select File** dialog box*

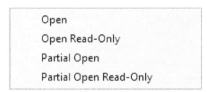

*Figure 1-37*  *The **Files of type** dialog box*

## Open Read-Only

To view a drawing without modifying it, select the **Open Read-Only** option from the drop-down list. If you try to save the opened drawing with the original file name, AutoCAD MEP warns you that the drawing file is write protected. However, you can save the edited drawing with a different file name using the **SAVEAS** command. This way you can preserve your drawing.

## Partial Open

The **Partial Open** option enables you to open only a selected view or a selected layer of a selected drawing. This option can be used to edit small portions of a complicated drawing and then save it with the complete drawing. When you select the **Partial Open** option from the **Open** drop-down list, the **Partial Open** dialog box is displayed, as shown in Figure 1-38, which contains different views and layers of the selected drawing. When you select a check box for a layer and then choose the **Open** button, only the objects drawn in that particular layer for the drawing are displayed in the new drawing window. You can make the changes and then save it.

*Figure 1-38  The **Partial Open** dialog box*

## Loading Additional Objects to Partially Opened Drawing

Once you have opened a part of a drawing and made the necessary changes, you may want to load additional objects or layers on the existing ones. To do so, enter **PARTIALOAD** at the Command prompt; the **Partial Load** dialog box will be displayed, which is similar to the **Partial Open** dialog box. You can choose another layer and the objects drawn in that layer will be added to the partially loaded drawing.

**Note**

*1. The **Partial Load** option is not enabled in the **File** menu unless a drawing is partially opened.*

*2. Loading a drawing partially is a good practice when you are working with objects on a specific layer in a large complicated drawing.*

*3. In the **Select File** dialog box, the preview of a drawing which was partially opened and then saved is not displayed.*

**Tip**. *If a drawing was partially opened and saved previously, it is possible to open it again with the same layers and views. AutoCAD MEP remembers the settings therefore when you open a partially opened drawing, a dialog box is displayed asking for an option to fully open it or restore the partially opened drawing.*

## Select Initial View

Select the **Select Initial View** check box if you want to load a specific view initially when AutoCAD MEP loads the drawing. This option will work, if the drawing has saved views. This is generally used while working on a large complicated drawing, in which you want to work on a particular portion of the drawing. You can save that particular portion as a view and then select it to open the drawing next time. You can save a desired view, by using the **VIEW** command. If the drawing has no saved views, then on selecting this option, the last view will be loaded. If you select the **Select Initial View** check box and then the **OK** button, AutoCAD MEP will display the **Select Initial View** dialog box. You can select the view name from this dialog box, and AutoCAD MEP will load the drawing with the selected view displayed.

# Opening an Existing Drawing Using the Startup Dialog Box

If you have set the **STARTUP** system variable value as **1**, the **Startup** dialog box will be displayed whenever you start a new AutoCAD MEP session. The first button in this dialog box is the **Open a Drawing** button. When you choose this button, a list of the most recently opened drawings will be displayed for you to select from, refer to Figure 1-39. The **Browse** button displays the **Select File** dialog box, which allows you to browse to another file.

*Figure 1-39  List of the recently opened drawings*

**Note**

*The display of the dialog boxes related to opening and saving drawings will be disabled, if the **STARTUP** and the **FILEDIA** system variables are set to 0.*

## Opening an Existing Drawing Using the Drag and Drop Method

You can also open an existing drawing in AutoCAD MEP by dragging it from the Window Explorer and dropping it into AutoCAD MEP. If you drop the selected drawing in the drawing area, the drawing will be inserted as a block and as a result you cannot modify it. But, if you drag the drawing from the Window Explorer and drop it anywhere other than the drawing area, AutoCAD MEP opens the selected drawing.

# QUITTING AutoCAD MEP

You can exit the AutoCAD MEP program by using the **EXIT** or **QUIT** commands. Even if you have an active command, you can choose **Exit AutoCAD MEP 2014** from the **Application Menu** to quit the AutoCAD MEP program. In case the drawing has not been saved, a dialog box is displayed with the **Yes** and **No** buttons. Choose the **Yes** button to save the drawing. Note that if you choose **No** in this dialog box, all the changes made in the current list till the last save will be lost. You can also use the **Close** button (**X**) of the main AutoCAD MEP window to end the AutoCAD MEP session.

# CREATING AND MANAGING WORKSPACES

A workspace is defined as a customized arrangement of **Ribbon**, toolbars, menus, and window palettes in the AutoCAD MEP environment. You can create your own workspaces, in which only specified toolbars, menus, and palettes are available. When you start AutoCAD MEP, by default, the **HVAC** workspace is the current workspace. You can select any other predefined workspace from the flyout displayed on choosing the **Workspace Switching** from the **Application Status Bar**, refer to Figure 1-40. You can also set the workspace by using the WORKSPACE command.

*Figure 1-40  The predefined workspaces*

## Creating a New Workspace

To create a new workspace, customize the **Ribbon** and invoke the palettes to be displayed in the new workspace. Next, select the **Save Current As** option from the flyout displayed on choosing the **Workspace Switching** button from the **Application Status Bar**; the **Save Workspace** dialog box will be displayed, as shown in Figure 1-41. Enter the name of the new workspace in the **Name** edit box and choose the **Save** button.

*Figure 1-41*   *The* **Save Workspace** *dialog box*

The new workspace is now the current workspace and is added to the drop-down list in the title bar. Similarly, you can create workspaces based on your requirement and switch from one workspace to another by selecting the name from the flyout displayed on choosing the **Workspaces Switching** button from the **Application Status Bar**.

## Modifying the Workspace Settings

AutoCAD MEP allows you to modify the workspace settings. To do so, select the **Workspace Settings** option from the flyout displayed on choosing the **Workspaces Switching** button from the **Application Status Bar**; the **Workspace Settings** dialog box will be displayed, as shown in Figure 1-42. All workspaces are listed in the **My Workspace** drop-down list. You can make any of the workspaces as My Workspace by selecting it in the **My Workspace** drop-down list. The options in the **Workspace Settings** dialog box are discussed next.

### Menu Display and Order Area

The options in this area are used to control the display and the order of display of workspaces in the **Workspace Switching** drop-down list. By default, workspaces are listed in the sequence of their creation. To change the order, select a workspace and choose the **Move Up** or **Move Down** button. To control the display of the workspaces, you can select or clear the check boxes. You can also add a separator between workspaces by choosing the **Add Separator** button. A separator is a line that is placed between two workspaces in the flyout displayed on choosing the **Workspaces Switching** button from the **Application Status Bar**, as shown in Figure 1-43.

*Figure 1-42*   *The* **Workspace Settings** *dialog box*

*Figure 1-43*   *The flyout displayed on choosing the* **Workspaces Switching** *button*

## When Switching Workspaces Area

By default, the **Automatically save workspace changes** radio button is selected in this area, so the changes made in the current workspace will be automatically saved when you switch to the other workspace. If you select the **Do not save changes to workspace** radio button then while switching the workspaces, the changes made in the current workspace will not be saved. Therefore, when you invoke this workspace again, it will display with default settings

# AutoCAD MEP'S HELP

You can get the on-line help and documentation about the working of AutoCAD MEP 2014 commands by using the options from the **Help** menu in the title bar, refer to Figure 1-44. You can access AutoCAD MEP's help by pressing the F1 function key. On pressing the F1 function key, the **AutoCAD MEP 2014 - Help** will be displayed, as shown in Figure 1-45. The entire help documentation on AutoCAD MEP 2014 is available on this page. You can search for information about any command or tool on this page. You can access this page without the internet connection also. But, if you want to use the offline help then you need to download the help file first. You can download the offline help from Autodesk website. Some important options in the **Help** menu are discussed next.

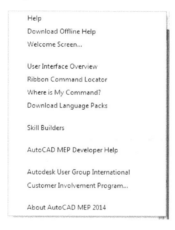

*Figure 1-44 The **Help** menu*

## Customer Involvement Program

This option is used to share information about your system configuration and uses of Autodesk products. The collective information is used by Autodesk for the improvement of Autodesk software.

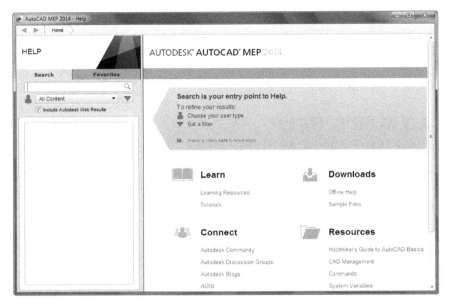

*Figure 1-45  The AutoCAD MEP 2014 - Help*

### About AutoCAD MEP 2014

This option gives you information about the Release, Serial number, Licensed to, and also the legal description about AutoCAD MEP.

# Autodesk Exchange Apps

Autodesk Exchange Apps enables you to customize AutoCAD MEP 2014 according to your requirements. In this window, you get various unique and design specific apps. These apps allow you to connect to the AutoCAD MEP network, share information and designs, and so on. On choosing the **Autodesk Exchange Apps** button from the title bar, the **Autodesk Exchange|Apps** window will be displayed in your default browser, refer to Figure 1-46. You can download and use any of the apps depending on your requirement.

# InfoCenter BAR

An **InfoCenter** bar is displayed at the top right corner in the title bar that will help you sign into the Autodesk Online services, refer to Figure 1-47. You can also access AutoCAD MEP community by using certain keywords.

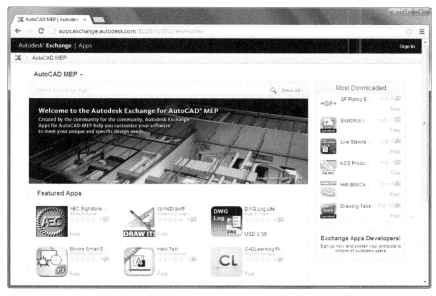

*Figure 1-46* *The **Autodesk Exchange | Apps** window*

*Figure 1-47* *The **InfoCenter** bar*

# ADDITIONAL HELP RESOURCES

1. You can get help for a command by pressing the F1 key while working. The help html containing information about the command is displayed. You can exit the dialog box and continue with the command.

2. You can get help about a dialog box by choosing the **Help** button in that dialog box.

3. Autodesk has provided several resources that you can use to get assistance with your AutoCAD MEP questions. The following is a list of some of the resources:

   a. Autodesk website *http://www.autodesk.com*
   b. AutoCAD MEP Technical Assistance website *http://www.autodesk.com/support*
   c. AutoCAD MEP Discussion Groups website *http://discussion.autodesk.com/index.jspa*

4. You can also get help by contacting the author, Sham Tickoo, at *techsupport@cadcim.com, tickoo525@gmail.com, and Stickoo@purduecal.edu*

5. You can download AutoCAD MEP drawings, programs, and special topics by visiting *www.cadcim.com*.

## Self-Evaluation Test

**Answer the following questions and then compare them to those given at the end of this chapter:**

1. You can increase the display area of the screen by using the **Clean Screen** button available in the **Application Status Bar**. (T/F)

2. The **AutoCAD Text window** is used to write text in the drawing area. (T/F)

3. The **Isolate Objects** button is used to display or hide the selected object. (T/F)

4. The **CLOSE** command is used to close the AutoCAD MEP application. (T/F)

5. To restrict the movement of cursor along a specific angle, you need to turn on the _____ snap.

6. Using the _____command, you can change the time interval of automatic save.

7. The **Startup** dialog box is displayed when the **STARTUP** variable is set to _____.

8. You can sign into the Autodesk Online Services by using the options available in the _____ bar.

**Answers to Self-Evaluation Test**
**1.** F, **2.** F, **3.** T, **4.** F, **5.** Polar Tracking, **6.** SAVETIME, **7.** 1, **8.** InfoCenter

# Chapter 2

# Getting Started with AutoCAD MEP

## Learning Objectives

*After completing this chapter, you will be able to:*

- *Understand the Workflow Path*
- *Use the Project Browser*
- *Understand the concept of Space*
- *Specify the Space Object Settings*
- *Work with Space Styles and Tools*
- *Understand the concept of zones*
- *Invoke different Workspaces in MEP*

# INTRODUCTION
AutoCAD MEP, a software based on AutoCAD platform, is used to design, draft, and document electrical, mechanical, and piping system of buildings. The first step while creating such a system is to create a project. Thereafter drawings are added to this project to represent various components of the building like electrical system, ducts, and so on. These drawings are then arranged according to the workflow of the project.

# WORKFLOW
Workflow is a sequence of connected steps required to create a specific type of system. It is dependent on the system to be created. For example, for an HVAC system, the workflow is shown in Figure 2-1. The steps involved in this workflow are discussed next.

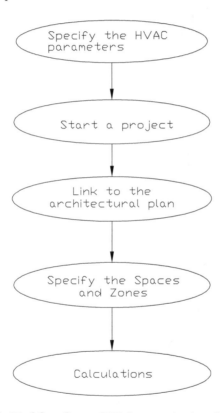

*Figure 2-1* *Workflow for an HVAC system in AutoCAD MEP*

## Specifying the HVAC Parameters
In this step, you need to specify all the parameters required to create an HVAC system. Some of the parameters are duct size, duct rise and drop, flow rate, cut length, and so on. On the basis of these parameters, the hierarchy of drawings in project will be decided.

## Starting a Project

In this step, you need to create a project file to link the drawings of a system to each other. For example, the drawings related to HVAC system will be interlinked and will be in the same project. As a result, you can easily manipulate the parameters of the system.

## Linking System File to the Architectural Plan

To create an HVAC system, you need to have an architectural plan. In this step, you will link the HVAC system file to the architectural plan file using the **Project Navigator**. After the completion of this step, you can reroute the HVAC lines according to the architectural plan.

## Specifying Spaces and Zones

In this step, you need to create spaces and zones using the tools available in AutoCAD MEP. These spaces are used for exporting the building information related to the heating and cooling loads. The spaces are further divided into zones which represent the actual heating and cooling loads.

## Calculating Loads

In this step, you need to calculate the heating and cooling loads for the building. These calculations are performed by using the analysis tools available in AutoCAD MEP.

**Note**
*In this textbook, the global unit system is followed, so you need to start AutoCAD MEP 2014 by double-click on the AutoCAD MEP 2014 - English (Global) icon from the desktop.*

# PROJECT BROWSER

The **Project Browser** is used to manage project files. A project file contains the record of drawing files related to a category. For example, a building project file may have a record of drawing files related to piping, electrical, ground floor, and first floor. You can create a new project file by using the options available in the **Project Browser**. Also, you can configure an existing project file. The **Project Browser** can be invoked by entering the **PROJECTBROWSER** command at the command prompt or by using the **Application** menu. To invoke it from the **Application** menu, hover the cursor on the **Open** option in the menu; a flyout will be displayed. Choose the **Project** button from the flyout; the **Project Browser** will be displayed, as shown in Figure 2-2.

In the **Project Browser**, the list of projects available in the selected directory is displayed in the left pane of the **Project Browser**. To change the current directory, choose the **My Projects** button from the left area of the dialog box; a drop-down list will be displayed having shortcuts to some common directories such as Documents and Desktop. Using these shortcuts, you browse to the desired directory and select the category as the current directory; all the projects available in the selected directory will be displayed in the left pane of the **Project Browser**. In addition to the existing projects, you can also create new project file by using the options available in the **Project Browser**. The procedure to create a new project file is discussed next.

# Creating a New Project File

A new project file can be created by using the **New Project** tool available at the bottom left corner of the **Project Browser**. On choosing this tool, the **Add Project** dialog box will be displayed, as shown in Figure 2-3.

*Figure 2-2  The* ***Project Browser***

*Figure 2-3  The* ***Add Project*** *dialog box*

There are three edit boxes available in this dialog box namely: **Project Number**, **Project Name**, and **Project Description**. The **Project Number** edit box is used to specify a unique number for the project file. The **Project Name** edit box is used to specify the name for the project file. The **Project Description** edit box is used to specify the description about the project file. You can use any of standard templates for the project. To do so, select the **Create from template**

**project** check box from the **Add Project** dialog box; the edit box will be activated below the check box. Next, choose the **Browse** button available next to the edit box activated; the **Select Project** dialog box will be displayed, as shown in Figure 2-4.

*Figure 2-4* *The* **Select Project** *dialog box*

Browse to the location where template files are stores and choose the open button after selecting the desired template file; the selected template file will be used for the current project. Note that in this textbook, the standard template used is *Commercial Template Project (Metric).apj*. To select the template, Browse to the location *C:\ProgramData\Autodesk\MEP 2014\enu\Template\ Commercial Template Project (Metric)* and then select the *Commercial Template Project (Metric)* template file. Next, choose the **Open** button; the path of the selected file will be added to the edit box displayed below the **Create from template project** check box. Specify the name of project in the **Project Name** edit box and choose the **OK** button from the **Add Project** dialog box; a project file with the specified name will be created and also the copy of drawing files in the template project will be created in the new project folder. The extension of the project file is *.apt*. Now, close the **Project Browser** by choosing the **Close** button. On closing the **Project Browser**; the **Project Navigator** will be displayed, refer to Figure 2-5.

# PROJECT NAVIGATOR

The **Project Navigator** is used to navigate the drawing files of a project. Using the **Project Navigator**, you can edit or create building drawings and other documentation data. The **Project Navigator** has four tabs: **Project**, **Constructs**, **Views**, and **Sheets**. These tabs are discussed next.

## Project Tab

The **Project** tab has three areas. These areas display the top level information of the project. The tools available at the bottom of the **Project Navigator** are used to configure the project. The options available in the **Project** tab are discussed next.

## Current Project
The **Current Project** area is used to store the project file name, project file number, and description of the project.

## Levels
The **Levels** area contains information of various levels used in the project file. In case of AutoCAD MEP level refers to floor. In this area, level names are displayed in the **Name** column and their corresponding height values are displayed in the **Elevation** column.

## Divisions Area
The **Divisions** area contains information of various divisions of the building. A division is the segment of a building along the horizontal plane.

## Project Browser
 The **Project Browser** tool is used to invoke the **Project Browser** dialog box. This dialog box has already been discussed.

## Close Current Project
The **Close Current Project** tool is used to close the current project file as well as the **Project Navigator**. The tool palette opened in the current project will also get closed.

## Content Browser
The **Content Browser** tool is used to display the **Autodesk Content Browser 2014** window related to the currently loaded project, refer to Figure 2-6. In this figure, the name of the current project is **Sample Project - Project Standards**. When you double-click on the project icon in the **Autodesk Content Browser 2014** window, different components related to the current project are displayed in different categories, refer to Figure 2-7.

*Figure 2-5  The **Project Navigator***

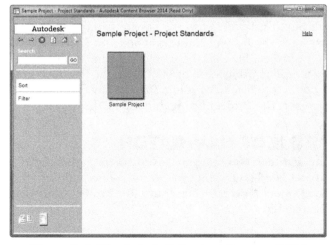

*Figure 2-6  The **Autodesk Content Browser 2014***

When you click on any of the categories shown in Figure 2-7; the components related to that category will be displayed in **Autodesk Content Browser 2014**. You can import any of the components available in **Autodesk Content Browser 2014** to the **Tool Palettes**. To do so, press and hold the left mouse button on the **i** icon of the component and then drag it to the **Tool Palettes**, refer to Figure 2-8. The component will be placed in the **Tool Palettes**.

*Figure 2-7  The **Autodesk Content Browser** displaying component categories*

*Figure 2-8  Component being placed in the **Tool Palettes***

## Refresh Project

The **Refresh Project** tool is used to refresh the current project. When a project file is used simultaneously by multiple individuals on a network, then those individuals may not be having the latest updated drawing. In such cases, it becomes very important to update the project file by using the **Refresh Project** button.

## Synchronize Projects

The **Synchronize Projects** tool is used to synchronize the current project with the AEC project standards. To synchronize the project, choose the **Synchronize Projects** tool available at the bottom of the **Project Navigator**; the **Synchronize Project with Project Standards** dialog box will be displayed, as shown in Figure 2-9. The drawing files available in the current project are displayed in the **Host Drawing** column of this dialog box. For every file, the status for the availability of the updated version is displayed in the **Status** column of the dialog box. You can synchronize the drawing files which show **Newer Version** in their **Status** column. To synchronize a drawing file, click on its corresponding field in the **Action** column; a drop-down list will be displayed. Select the **Update from Standard** option from the drop-down list; the selected file will be synchronized. Similarly, you can synchronize more than one file at a time by pressing and holding the CTRL key while selecting the files. After selecting the **Update from Standard** option for all the selected files.

*Figure 2-9 The Synchronize Project with Project Standards dialog box*

## Configure Project Standards

The **Configure Project Standards** tool is used to set up the project standard. These project standards are used to synchronize the project drawings. When you choose this tool, the **Configure AEC Project Standards** dialog box will be displayed. This dialog box has already been discussed in Chapter 1.

# Constructs Tab

The options in the **Constructs** tab are used to modify the drawing files available in the project. On choosing the **Constructs** tab, the **Project Navigator** will be modified, as shown in Figure 2-10. Some of the options in this tab have already been discussed. The remaining options available in this tab are discussed next.

## Constructs Area

The **Constructs** area has all the drawings of the current project arranged in a tree structure. In the tree, the drawing files are divided into two main categories: **Constructs** and **Elements**, which are further categorized according to their purpose/function. For example, the architectural drawings are stored in the Architectural category. You can open any of the drawings available in these folders by double-clicking on it.

## Add Category

The **Add Category** tool is used to add a new category in the **Project Navigator** to categorize the drawing files according to their usage. To do so, select a node from the **Constructs** rollout of the **Project Navigator** and then choose the **Add Category** tool; the category will be added under the selected node.

## Add Construct

The **Add Construct** tool is used to add a new construction in the project. When you choose this tool, the **Add Construct** dialog box will be displayed, as shown in Figure 2-11. You can add description about the new construct, change the name, edit category, and change the template using the fields available in the dialog box. In this dialog box, check boxes corresponding to each level of floor in the building are available in the **Assignments** area. Select the check boxes corresponding to the levels to which you want to add the construction. If you select multiple check boxes in this dialog box, then the objects created in the construction will span between the levels selected. Therefore, it is recommended to select only one check box. You can open the newly created construct by using the **Open in drawing editor** check box. When you select this check box and choose the **OK** button, the **Add Construct** dialog box will close and the newly created construction will open in AutoCAD MEP for editing.

*Figure 2-10* The *Project Navigator* with the *Constructs* tab chosen

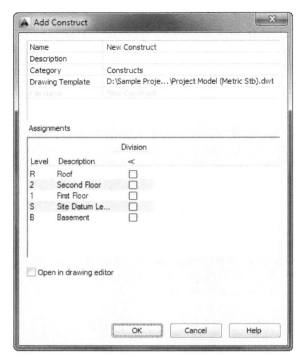

*Figure 2-11  The **Add Construct** dialog box*

## Add Element

The **Add Element** tool is used to add a new element to the project. When you choose this tool, the **Add Element** dialog box will be displayed. Specify the required parameters and choose the **OK** button; a new element is added to the **Elements** category. You can edit the newly added element by selecting the **Open in drawing editor** check box.

## Show External Reference

The **Show External Reference** tool is used to display external references for any drawing in the current project. When you select a drawing in the **Project Navigator** and choose this button, the **External References** dialog box will be displayed, refer to Figure 2-12. In the **Details** area of this dialog box, the information about the selected drawing is displayed.

## Views Tab

On choosing this tab, the **Project Navigator** will be modified, refer to Figure 2-13. The options in the **Views** tab are used to add, modify, or delete a general view, a detail view, or a section view in the project. Some of the options in this tab have already been discussed. The remaining options in this tab are discussed next.

## Views Area

In this area, different views of the project drawings are displayed in a tree structure. These views are divided into different categories on the basis of their application areas. For example, all the drawings related to interior of the building are available in the **Interiors** category in the **Views** tree. You can open any of the drawing views by double-clicking on it.

*Figure 2-12* The **External References** *dialog box*

*Figure 2-13* The **Project Navigator** *with the* **Views** *tab chosen*

## Add View

There are three types of views that can be added to the **View** area: General view, Section/ Elevation view, and Detail view. The **Add View** tool is used to add a new view to the **View** area. On choosing the **Add View** tool, the **Add View** dialog box will be displayed, as shown in Figure 2-14.

*Figure 2-14* The **Add View** *dialog box*

There are three options available in this dialog box: **Add a General View**, **Add a Section/ Elevation View**, and **Add a Detail View**. On choosing an option, the respective dialog box will be displayed. For example, when you choose the **Add a General View** option, the **Add General View** dialog box will be displayed, as shown in Figure 2-15

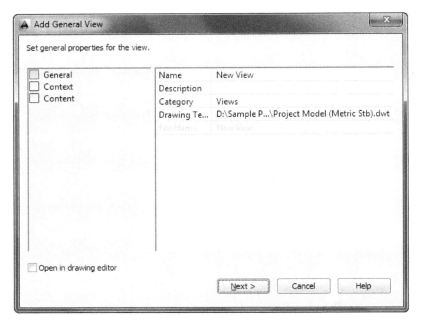

*Figure 2-15  The **Add General View** dialog box*

Using the options available in this dialog box, you can set the general properties of the view. Three pages are available in this dialog box, **General**, **Context**, and **Content**. The **General** page is displayed by default in this dialog box. In this page, you need to enter the general informations about the view. After entering the general information, choose the **Next** button; the **Context** page will be displayed, refer to Figure 2-16.

*Figure 2-16  The **Add General View** dialog box with the **Context** page displayed*

On this page, you can select the portions of the building that you want to include in the current view. To do so, you can select the check boxes corresponding to the portions of the building required to be included in the current view. After selecting the check boxes, choose the **Next** button; the **Content** page will be displayed, refer to Figure 2-17.

*Figure 2-17  The **Add General View** dialog box with the **Content** page displayed*

In this page, you can select the check boxes corresponding to those elements that you want to display in the current view. After specifying the required parameters, choose the **Finish** button to exit the dialog box. The newly created view will be displayed in the **Views** area of the **Project Navigator**.

## Regenerate View

The **Regenerate View** tool is used to regenerate all the views so that you can get the updated version of the drawing views. This tool is also used to update the reference of the views.

## Repath Xref

The **Repath Xref** tool is used to reconnect the external references whose names or locations are changed.

# Sheets Tab

On choosing this tab, the **Project Navigator** will be modified, refer to Figure 2-18. The options in the **Sheets** tab are used to add, modify, or delete the sheets available in the project. Some of the options in this dialog box have already been discussed. The remaining options are discussed next.

*Figure 2-18* The **Project Navigator** *with the* **Sheets** *tab chosen*

## Sheet Set View Area
All the drawing sheets available in the current project are displayed in the **Sheet Set View** area in a tree structure. You can open any of the drawing sheets by double-clicking on it.

## Add Sheets
The **Add Sheets** tool is used to add a new sheet in the project. To do so, choose this tool, the **New Sheet** dialog box will be displayed, as shown in Figure 2-19. In this dialog box, you need to specify the sheet number, sheet title, and file name in their respective edit boxes. On specifying the sheet number and the sheet title in the edit boxes, a default file name is displayed in the **File name** edit box. You can change this file name or retain it. The folder path and the template path for the current sheet are displayed in the **Folder path** and **Sheet template** fields, respectively. These fields cannot be modified in this dialog box. After specifying the required parameters, choose the **OK** button from this dialog box; the newly created sheet will be added in the project and will be displayed at the bottom of the list in the **Project Navigator**. You can edit this sheet by double-clicking on it.

*Figure 2-19* The **New Sheet** *dialog box*

## Publish

The **Publish** tool is used to publish the complete sheet set in DWF, PDF, or DWFx file format. To publish a sheet set, choose the **Publish** tool from the **Project Navigator**; the **AutoCAD MEP 2014** message box will be displayed, as shown in Figure 2-20. Choose the **OK** button from this message box; the **Specify DWFx File** dialog box will be displayed, as shown in Figure 2-21. Choose the **Select** button from this dialog box; the **Publish Job Progress** message box will be displayed.

*Figure 2-20  The **AutoCAD MEP 2014** message box*

This message box shows the progress of plotting/publishing of the drawing sheets. When the process of publishing completes, the **Plot and Publish Job Complete** message will be displayed at the bottom right corner and the plot file will be created at the location specified in the **Specify DWFx File** dialog box. To view the details of the plot file, click on the **Click to view plot and publish details** link from the message box displayed at the bottom right corner of the application window; the **Plot and Publish Details** dialog box will be displayed with the details of the plot file.

*Figure 2-21  The **Specify DWFx File** dialog box*

# SPACE

Space is an entity used in AutoCAD MEP to find out technical information about a specific section of an architectural drawing of a project. For example, if you want to make a specific section of a drawing as an office, then you need to assign corresponding space style to that section. The properties of that office such as area, airflow, height, equipment, and load are stored in the space style. To assign a space style to an area, you need to create a space first. The method to create a space is discussed next.

## Creating Spaces

To create a space, choose the **Space** tool from the **Space** drop-down in the **Build** panel of the **Home** tab of the **Architecture** workspace. This tool is also available in the other workspaces. You can invoke this tool from any of the workspaces by entering the **SPACEADD** command at the Command prompt. On doing so, the **Properties** palette will be displayed and also a space of default size will be attached to the cursor. You need to click in the drawing area to specify the position of the space. As soon as you specify the position of space, you will be prompted to specify the rotation value for the space. Specify the rotation value; the space will be positioned at the defined point. The Command prompt for creating the space is given next.

*Choose the* **Space** *tool.*
Start corner or [Name/STyle/Create type/Height]: **C**
Set create type [Insert/Polygon/Generate] <Rectangle>: **I**
Insertion point or [Name/STyle/Create type/Length/WIdth/Height/MOve/SIze/Drag point/MAtch]: *Enter the co-ordinates of the insertion point or click in the drawing area.*
Rotation or [Name/STyle/Create type/Length/WIdth/Height/MOve/SIze/Drag point/MAtch/Undo] <0>: *Enter the rotation angle of the space.*
Insertion point or [Name/STyle/Create type/Length/WIdth/Height/MOve/SIze/Drag point/MAtch/Undo]:

The options available in the Command prompt are discussed next.

### Name

The **Name** option is used to define a name for the space created by using the **SPACEADD** command. By default, the **Space** name is assigned to the newly created space.

### STyle

The **STyle** option is used to change the style of the space. By default, the **Standard** style is selected. You can change the space style by using the **SPACESTYLE** command. This command will be discussed later in this chapter.

### Create type

The **Create type** option is used to change the type of space to be created. The space to be created can be a rectangle, a polygon, or it can be associative to the boundary objects. The Command prompt after selecting the **Create type** option is given next.

Start corner or [Name/STyle/Create type/Height]: **C**
Set create type [Rectangle/Polygon/Generate] <Insert>:

According to the above Command prompt, there are four options to create a space. These options are discussed next.

**Insert:** By default, this option is selected at the Command prompt. Press the ENTER key; the space will be created according to the target dimensions specified in the design rules of the space style. Figure 2-22 shows a space created by using the **Insert** option. The Command prompt for creating a space by using the **Insert** option is given next.

*Figure 2-22* *The space created by using the* ***Insert*** *option*

Insertion point or [Name/STyle/Create type/Length/WIdth/Height/MOve/SIze/Drag point/MAtch]: *Specify the insertion point.*
Rotation or [Name/STyle/Create type/Length/WIdth/Height/MOve/SIze/Drag point/MAtch/Undo] <default value>: *Specify the rotation value for the space.*
Insertion point or [Name/STyle/Create type/Length/WIdth/Height/MOve/SIze/Drag point/MAtch/Undo]:

**Rectangle:** By choosing this option, you can specify the length and width of the space to be inserted in the drawing area. When you choose this option and specify the first corner point, the other corner point of the rectangle will get attached to the cursor, as shown in Figure 2-23. Also, you will be prompted to specify the end corner point. Either click in the drawing area to specify the end corner point or enter the co-ordinates of the end corner point.

*Figure 2-23* *The rectangular space with cursor attached to the corner point of the rectangle*

**Polygon:** By choosing this option, you can create a polygonal space by manually defining the segments of the polygon. You can create a polygon type space either by using the arcs or by using the lines. Figure 2-24 shows the polygon type space created by using the arcs and Figure 2-25 shows the polygon type space created by using the lines.

*Figure 2-24  The polygon type space created by using arcs*

*Figure 2-25  The polygon type space created by using lines*

**Generate:** On choosing this option, the space created will be associative to the boundary objects. If you do not have any object with its space boundaries defined or with a valid space boundary, then the **Analyzing Potential Spaces** dialog box will be displayed, as shown in Figure 2-26. You can use all the visible objects to create a bounding space by choosing the **Use all visible objects to bound spaces** option from the dialog box. You can also select an object to create the bound space. To do so, choose the **Select objects that should bound spaces** option from the dialog box.

*Figure 2-26  The Analyzing Potential Spaces dialog box*

## Length

The **Length** option is used to change the length of the space to be created. By default, the target dimension of the space style is used as the length of the space. The Command prompt to change the length is given next.

Insertion point or [Name/STyle/Create type/Length/WIdth/Height/MOve/SIze/Drag point/MAtch]: **L**
Length <default value>: *Enter the desired value of length.*

## WIdth

The **WIdth** option is used to change the width of the space to be created. By default, the target dimension of space style is used as the width of the space. The Command prompt to change the width is given next.

Insertion point or [Name/STyle/Create type/Length/WIdth/Height/MOve/SIze/Drag point/MAtch]: **WI**
Length <default value>: *Enter the desired value of width.*

## Height

The **Height** option is used to change the height of the space to be created. By default, the target dimension of space style is used as the height of the space. The Command prompt to change the height is given next.

Insertion point or [Name/STyle/Create type/Length/WIdth/Height/MOve/SIze/Drag point/MAtch]: **H**
Length <default value>: *Enter the desired value of height.*

## MOve

The **MOve** option is used to move the space created. This option is available only if you have chosen **Insert** from the **Create type** options. After moving the space, you can also rotate the space by a specified angle. The Command prompt to move the space is given next.

Insertion point or [Name/STyle/Create type/Length/WIdth/Height/MOve/SIze/Drag point/MAtch]: **MO**
Insertion point or [Name/STyle/Create type/Length/WIdth/Height/MOve/SIze/Drag point/MAtch]: *Specify the insertion point by clicking or entering the coordinates.*
Rotation or [Name/STyle/Create type/Length/WIdth/Height/MOve/SIze/Drag point/MAtch/Undo] <0>: *Specify the rotation value by clicking or entering the coordinates.*

## SIze

The **SIze** option is used to specify the size of the space to be created. This option is available only if you have selected **Insert** from the **Create type** options. The Command prompt for specifying the size of the space is given next.

Insertion point or [Name/STyle/Create type/Length/WIdth/Height/MOve/SIze/Drag point/MAtch]: **SI**
Insertion point or [Name/STyle/Create type/Length/WIdth/Height/MOve/SIze/Drag point/MAtch]: *Specify the insertion point by clicking or entering the coordinates.*
New size or [Name/STyle/Create type/Length/WIdth/Height/MOve/SIze/Drag point/MAtch/Undo] <3000>: *Specify the insertion point by clicking or entering the dimensions.*
Rotation or [Name/STyle/Create type/Length/WIdth/Height/MOve/SIze/Drag point/MAtch/Undo] <0>: *Specify the rotation value by clicking or entering the coordinates.*

## Drag point

The **Drag point** option is used to change the orientation of the space to be created. This option is available only if you have selected **Insert** from the **Create type** options. The Command prompt to change the orientation of the space is given next.

Insertion point or [Name/STyle/Create type/Length/WIdth/Height/MOve/SIze/Drag point/MAtch]: **D**
Insertion point or [Name/STyle/Create type/Length/WIdth/Height/MOve/SIze/Drag point/MAtch]: *Enter D again if you want to change the orientation or press ENTER to exit.*

## MAtch

The **MAtch** option is used to match the style of the newly created space with an existing space. This option is available only if you have selected **Insert** from the **Create type** options. The Command prompt to change the orientation of the space is given next.

Insertion point or [Name/STyle/Create type/Length/WIdth/Height/MOve/SIze/Drag point/MAtch]: **MA**
Select a space to match: *Select a space created earlier to use its properties for new space.*
Match [Style/Length/Width/Height] <All>: *Enter any of the options available in the prompt to match the properties. By default, the **All** option is selected, so all the properties of selected space are copied in the new space. Now, click in the drawing area to place it.*

## Arc

This option is used to create an arc in a polygon type space. The **Arc** option is available only when **Polygon** is selected from the **Create type** options. The command prompt to create a polygon type space using an arc is discussed next.

Insertion point or [Name/STyle/Create type/Length/WIdth/Height/MOve/SIze/Drag point/MAtch]: **C**
Set create type [Rectangle/Polygon/Generate] <Insert>:**P**
Start point or [Name/STyle/Create type/Height/Arc]:**A**
Start point or [Name/STyle/Create type/Height/Line]: *Specify the first point of the arc.*
Second point or [Name/STyle/Create type/Height/Line/Undo]: *Specify the second point of the arc.*
Next point or [Name/STyle/Create type/Height/Line/Undo]: *Specify the third point to complete the arc.*
Second point or [Name/STyle/CReate type/Height/Line/Close/Ortho/Undo]: *Specify the next point or enter C to close the arc for creating the space.*

## Line

This option is used to create a polygon by using the line. The **Line** option is available only when **Polygon** is selected from the **Create type** options and then **Arc** is chosen in the next prompt. The Command prompt to create a polygon type space using lines is discussed next.

Insertion point or [Name/STyle/Create type/Length/WIdth/Height/MOve/SIze/Drag point/MAtch]: **C**

Set create type [Rectangle/Polygon/Generate] <Insert>: **P**
Start point or [Name/STyle/Create type/Height/Arc]: **A**
Start point or [Name/STyle/Create type/Height/Line]: **L**
Start point or [Name/STyle/Create type/Height/Arc]: *Specify the start point of the polygon.*
Next point or [Name/STyle/Create type/Height/Arc/Undo]: *Specify the end point of the first line of the polygon.*
Next point or [Name/STyle/Create type/Height/Arc/Ortho/Undo]: *Specify the end point of the second line of the polygon.*
Next point or [Name/STyle/CReate type/Height/Arc/Close/Ortho/Undo]: *Specify the next point or enter C to close the polygon.*

# Editing Spaces

You can perform various editing operations on the created spaces. For example, you can change the space style or divide a space. When you select a space from the drawing area, the **Space** contextual tab will be available in the **Ribbon**, refer to Figure 2-27. The options in this tab are discussed next.

*Figure 2-27  The **Space** contextual tab*

## Select Similar

The **Select Similar** tool is used to select all those components from the drawing area which have the same style and layer as the selected object. The objects with the same style and properties but different layers will not be selected by this tool.

## Object Viewer

The **Object Viewer** tool is used to display the selected object in a separate 3D preview window. When you choose this tool, the selected object will be displayed in a separate window, as shown in Figure 2-28.

Various tools such as **Parallel**, **Perspective**, and **Zoom Window** are in the main toolbar of this dialog box. These tools are used to change the display of view.

## Isolate Objects Drop-down

This drop-down is available in the **General** panel of the **Space** tab of the **Ribbon**. The tools in this drop-down are used to control the visibility of the objects. These tools are discussed next.

### Isolate Objects

This tool is used to hide all the deselected objects from the drawing area.

### Hide Objects

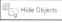
This tool is used to hide the selected objects from the drawing area.

### End Isolation

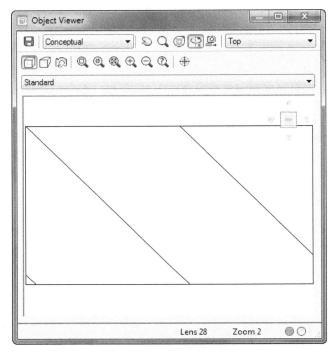 This tool is used to display all the hidden objects.

*Figure 2-28  The **Object Viewer** dialog box*

## Edit in Section

This drop-down is available in the **General** panel of the **Space** tab in the **Ribbon**. The tools in this drop-down are used to edit the space created. These tools are discussed next.

### Edit in Section

This tool is used to edit an object in a predefined section.

### Edit in Elevation

This tool is used to edit the space at a certain elevation distance from the selected reference.

### Edit in Plan

This tool is used to edit the space at certain plan distance from the selected reference.

## Add Selected

The **Add Selected** tool is available in the **General** panel of the **Space** tab in the **Ribbon**. This tool is used to add an instance of the selected object in the drawing area.

## Edit Style

This drop-down is available in the **General** panel of the **Space** tab in the **Ribbon**. The tools in this drop-down are used to modify the style of the space. These tools are discussed next.

### Edit Style

By using the options in this dialog box, you can change different properties of the selected space such as length and width of the space, target area, and various offset values. When you choose this tool, the **Space Style Properties** dialog box will be displayed, as shown in Figure 2-29. Using the options in the dialog box, you can modify the style of the selected space.

*Figure 2-29* *The Space Style Properties dialog box*

### Space Styles

When you choose this tool, the **Style Manager** dialog box will be displayed, as shown in Figure 2-30. The space styles available in the current drawing file are displayed in the left area of this dialog box under the **Space Styles** category. If you select any space style from this area, the options to change the space style will be displayed on the right area in the dialog box. These options are similar to the options displayed in the **Space Style Properties** dialog box.

### Display Theme Style

When you choose this tool, the **Style Manager** dialog box will be displayed, as shown in Figure 2-31. Various theme styles available in the drawing are displayed in the left area of this dialog box under the **Display Theme Styles** category. If you choose a theme style from this area, the options to change the properties of the selected theme are displayed in the right area. Using these options, you can change the properties of the selected theme like its title, text style, symbol, and so on.

*Figure 2-30* The **Style Manager** *dialog box*

*Figure 2-31* The **Style Manager** *dialog box with theme style options*

### Zone Templates

The **Zone Templates** tool is used to change the properties of the selected zone template such as its contents, modifiers, and name. To do so, choose the **Zone Templates** tool; the **Style Manager** dialog box will be displayed, as shown Figure 2-32.

*Figure 2-32 The Style Manager dialog box with the Zone Templates options*

Various zone template styles available in the drawing are displayed in the left area of this dialog box under the **Zone Templates** category. When you choose a template style from the left area, the options to change the properties of the selected template style are displayed in the right area. Using these options, you can change the properties of the selected template style.

Similarly, you can change the zone style by using the **Zone Style** tool available in the same drop-down.

## Copy Style

This tool is used to create a copy of the style of the selected object.

## Dimension

This tool is used to add AEC dimension to an object. To add a dimension, choose this tool and select the object to which you want to add dimension; the dimension will be attached to the cursor. Now, you can place the dimension at the desired location.

## Tag

This tool is used to add a tag to the object. To add a tag, choose **Space > General > Tag** from the **Ribbon**; you will be prompted to select an object on which you want to add a tag. On selecting the object, the tag will be attached to the cursor. Now, place the tag at the desired location. When you click to place the tag, the **Edit Property Set Data** dialog box will be displayed, as shown in Figure 2-33.

*Figure 2-33  The **Edit Property Set**
**Data** dialog box*

Using the fields available in this dialog box, you can change the properties of the tag for the selected object.

## Update Drop-down

This drop-down is available in the **Modify** panel. The tools available in this drop-down are used to update spaces. There are two tools available in this drop-down: **Selected Space** and **All Associative Spaces**. The **Selected Space** tool is used to update the selected space. The **All Associative Spaces** tool is used to update all the spaces associated with the selected space.

## Edit in Place

This tool is used to edit an extruded 3D space. To do so, choose **Space > Modify > Edit in Place** from the **Ribbon**; the **Edit in Place: Space Body Modifier** contextual tab will be displayed and the selected space will be converted into a free form space. Now, using the vertices and control points available in the free form space, you can edit the shape and size of the space. You can cycle between various options to edit a space by pressing the CTRL key after selecting the center point of its corresponding face.

**Note**
*While using the **Edit in Place** tool, when you hover the cursor over the center point of a selected face, an information box with the editing options will be attached to the cursor, refer to Figure 2-34.*

## Edit Surfaces

This tool is used to edit only the faces of the space. Choose **Space > Modify > Edit Surfaces** from the **Ribbon**, the selected space will be converted into a free form space. Now, using the midpoints of edges, you can edit the faces of the selected space. Also, a plus icon will be displayed at the bottom face of the space, refer to Figure 2-35. Using this plus icon, you can add windows and doors to the selected space.

*Figure 2-34* *The information box showing various editing options*

*Figure 2-35* *Space with editing points*

## Edit Vertices

This tool is used to add a vertex to the selected edge. To do so, choose **Space > Modify > Edit Vertices** from the **Ribbon**; the vertex will be attached to the cursor and you will be prompted to click in the drawing area to add the vertex, refer to Figure 2-36. Next, click in the drawing area; the vertex will be added to the corresponding edge. To remove the vertex, press and hold the CTRL or SHIFT key and then click on the vertex to be removed.

*Figure 2-36* *Space with its vertex attached to the cursor*

**Note**

*To use the **Edit in Place**, **Edit Surfaces**, and **Edit Vertices** tools, you need to switch to isometric viewport.*

## Make Associative

This tool is used to make the bounding objects associative to the selected space. To do so, choose **Space > Modify > Make Associative** from the **Ribbon**; the objects bounding the space will become associative with it. Note that for making space associative with the object, the boundary of the object must enclose the selected space.

## Create Polyline

This tool is used to create a polyline around the selected space. To create a polyline around a selected space, choose **Space > Modify > Create Polyline** from the **Ribbon**; a polyline will be created around the selected space.

## Divide Space

The **Divide Space** tool is available in the **Modify** panel of the **Space** tab. This tool is used to divide the selected space by using a line.

## Interference

The tools available in this drop-down are used to add or remove an interference condition from a selected space.

## Remove Void

The **Remove Void** tool is used to remove a selected void from the space.

## Space/Zone Manager

The **Space/Zone Manager** tool is available in the **Helpers** panel of the **Space** tab. This tool is used to change the properties of the selected space or zone.

## Show Boundaries

The **Show Boundaries** tool is available in the **Helpers** panel of the **Space** tab. This tool is used to show or hide the boundary of the selected space. This tool is available only for associated spaces.

# ZONE

A zone is a group of spaces which are used for a specific function; for example, a zone created for a specific temperature condition. You can create a zone or zones by using the **Zone** tool available in the **Space** drop-down of the **Build** panel in the **Home** tab of the **Ribbon**. When you choose this tool, a box will get attached to the cursor and you will be prompted to specify the location where the zone tag is to be inserted. Click in the drawing area; the zone will be created and name of the zone will be displayed along with the box. By default, **Zone** is displayed as the name for the created zone. You can change this name by using the **Name** field in the **Property** palette. You can edit a zone by using the options available in the **Zone** contextual tab. The **Zone** contextual tab is displayed on selecting the zone. The options available in this tab have already been discussed this chapter.

# WORKSPACES

A workspace is a combination of menus, toolbars, **Ribbon**, palettes, and control panels. It is used to represent a customized drawing environment based on the user requirement. In AutoCAD MEP, six default workspaces are available in the **Workspace Switching** flyout of the **Application Status bar**, refer to Figure 2-37. These workspaces are **HVAC**, **Piping**, **Electrical**, **Plumbing**, **Schematic**, and **Architecture**.

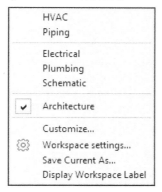

*Figure 2-37  The **Workspace Switching** flyout*

The options in the **Workspace Switching** flyout are discussed next.

## HVAC

The tools in the **HVAC** workspace are used for designing Heating, Ventilation, and Air Conditioning system. Therefore, on invoking the **HVAC** workspace, the tools related to the HVAC design will be displayed.

## Piping

The tools in the **Piping** workspace are used for routing and creating pipe lines in the building. To switch to this workspace, choose the **Piping** option from the **Workspace Switching** flyout; the options in the **Ribbon** and **Tool Palettes** will change according to the workspace. Note that some options for plumbing are also available in this workspace.

## Electrical

The tools in the **Electrical** workspace are used for creating electrical circuits, panels, devices, equipment, and cable trays. To switch to this workspace, choose the **Electrical** option from the **Workspace Switching** flyout; the options in the **Ribbon** and **Tool Palettes** will change according to the workspace.

## Plumbing

Tools available in this workspace are similar to the tools in the **Piping** workspace. The tools in this workspace are used for routing and creating plumbing lines in the building. To switch to this workspace, choose the **Plumbing** option from the **Workspace Switching** flyout; the options in the **Ribbon** and **Tool Palettes** will change according to the workspace. Some of the options for piping are also available in this workspace. Plumbing is concerned with the drainage and other water related applications, whereas piping has a broad application area such as gas supply and petroleum lines.

## Schematic

Schematics is the symbolic representation of an object. Using the **Schematic** workspace, you can represent MEP objects in the form of symbols. Some of the tools available in this workspace are schematic lines and schematic symbols.

## Architecture

The tools in the **Architecture** workspace are used to create architectural objects such as walls, doors, windows, and stairs. An architectural layout creates foundation for other domains like piping, HVAC, electrical, and so on. To switch to this workspace, choose the **Architecture** option from the **Workspace Switching** flyout; the **Ribbon** and **Tool Palettes** will change accordingly.

You can switch between the workspaces anytime during designing by using the **Workspace Switching** flyout. Also, you can customize any of the workspaces by using the **Customize** tool available in the **Workspace Switching** flyout. This tool is discussed next.

## Customize

The **Customize** tool is used to customize the user interface of AutoCAD MEP. On invoking this tool from the **Workspace Switching** flyout, the **Customize User Interface** dialog box will be displayed, as shown in Figure 2-38. The areas in this dialog box are discussed next.

*Figure 2-38  The **Customize User Interface** dialog box*

## Customization in All Files

The options in this area are used to select the customization file to be used for modification. Also, you can change the properties of the selected item from this area. On selecting an option from this area, the related properties are displayed in the **Properties** area available on the right of this dialog box.

## Command List

The options in this area are used to customize the selected command. On selecting a command from this area, the **Button Image** area is displayed at the top right corner of this dialog box. You can change the icon used for the selected button by using the options available in this area.

## Properties

The options in this area are used to change the specifications of the selected option.

## Self-Evaluation Test

**Answer the following questions and then compare them to those given at the end of this chapter:**

1. The extension for AutoCAD MEP project files is *.apt*. (T/F)

2. While changing the current directory of project files, you can create a category. (T/F)

3. The **Project Navigator** is used to manage the drawing files of a project. (T/F)

4. In AutoCAD MEP, a level refers to a _____ .

5. The _____ tool is used to display the **Autodesk Content Browser** related to the current loaded project.

6. A zone is a division of _____ which is used for a specific function.

7. The _____ tool is used to add an instance of the selected object to the drawing.

## Review Questions

**Answer the following questions:**

1.  The **Synchronize Projects** tool is used to synchronize the current project with the previous project standards. (T/F)

2.  You can create a space by using the **SPACEADD** command. (T/F)

3.  The **Customize** tool is used to customize the user interface of AutoCAD MEP. (T/F)

4.  Schematics is the _____ representation of an object.

5.  You can remove a vertex by pressing and holding the _____ or _____ key and clicking on the vertex.

# Chapter 3

# Working with Architecture Workspace

## Learning Objectives

*After completing this chapter, you will be able to:*

- *Add columns and grids*
- *Add walls*
- *Add windows and doors*
- *Add stairs, railings, and stair tower*
- *Add roofs and slabs*
- *Create layouts*

# INTRODUCTION

In this chapter, you will learn to work with the **Architecture** workspace. Before proceeding, it is important to understand that architecture is the process of planning, designing and constructing buildings and other structures. The architectural operations also include the architectural plan. An architectural plan is the sketch based on a set of conventional parameters such as sheet size, views, and units. Advancements in CAD software have not just influenced its wide area of application but have also enabled the architects to improve their architectural operations. AutoCAD MEP is one such CAD software that is used for creating or modifying these architectural plans.

# ARCHITECTURE WORKSPACE

To work with the architectural plans, you need to invoke the **Architecture** workspace. To do so, choose the **Architecture** option from the flyout displayed on choosing the **Workspace Switching** button available in the **Application Status** bar; the tools required for performing architectural operations will be displayed in the **Ribbon**, refer to Figure 3-1.

*Figure 3-1　The **Ribbon** displayed on invoking the **Architecture** workspace*

The procedure of creating various components of an architectural structure such as wall, door, window, grid, and so on are discussed next.

# CREATING WALLS

Walls are the building blocks of a structure. The tools to create a wall are available in the **Wall** drop-down of the **Build** panel in the **Home** tab of the **Ribbon**, refer to Figure 3-2. There are three tools for creating walls: **Wall**, **Curtain Wall**, and **Curtain Wall Unit**. These tools are discussed next.

*Figure 3-2　The **Wall** drop-down*

## Wall

This tool is used to create straight walls and curved walls. To create a wall, choose the **Wall** tool from the **Wall** drop-down available in the **Build** panel of the **Home** tab in the **Ribbon**; you will be prompted to specify the start point of the wall. Also, the **Properties** palette will be displayed. The prompt sequence for creating a wall by choosing the **Wall** tool is given next.

Start point or [STyle/Group/WIdth/Height/OFfset/Flip/Justify/Match/Arc]: *Specify the Start point of the wall segment.*
End point or [STyle/Group/WIdth/Height/OFfset/Flip/Justify/Match/Arc]: *Specify the end point of the current wall segment and starting point of the next wall segment.*
End point or [STyle/Group/WIdth/Height/OFfset/Flip/Justify/Match/Arc/Undo]: *Specify the end point of the current wall segment and starting point of the next wall segment.*
End point or
[STyle/Group/WIdth/Height/OFfset/Flip/Justify/Match/Arc/Undo/Close/ORtho close]: *Specify*

*the end point of the current wall segment and starting point of the next wall segment.*
End point or
[STyle/Group/WIdth/Height/OFfset/Flip/Justify/Match/Arc/Undo/Close/ORtho
close]: Enter

**Note**
*You can also invoke the tools given in the **Ribbon** by entering the corresponding command at the Command prompt.*

There are various options available in the command prompt to change the properties of the wall. Alternatively, you can select these options from the shortcut menu displayed on right-clicking in the drawing area. These options are discussed next.

## STyle

This option is used to change the style of the wall to be created. You can create a wall style by using the **Style Manager** which will be discussed later in this chapter. By default, the **Standard** style is chosen as the wall style.

## Group

This option is used to specify the cleanup group for the wall. These cleanup groups are creating by using **Style Manager** which is discussed later. When two walls of the same group intersect each other, they get automatically cleaned up at the intersection point, refer to Figure 3-3. Figure 3-4 shows the intersecting walls of two different groups.

***Figure 3-3*** *The intersecting walls of same cleanup group*

***Figure 3-4*** *The intersecting walls created of two different groups*

## WIdth

This option is used to specify the width of a wall. The width specified using this option will be applicable for all the walls that are created after specifying this value. The prompt sequence for changing the width of a wall is given next.

Start point or [STyle/Group/WIdth/Height/OFfset/Flip/Justify/Match/Arc]: **WI**
Width <254>: *Specify the desired width.*
Start point or [STyle/Group/WIdth/Height/OFfset/Flip/Justify/Match/Arc]: Enter

## Height

This option is used to specify the height of the wall. The height specified using this option will be used for all the walls to be created afterwards. To view the height of wall, you need to switch to the isometric view. The prompt sequence for changing the height of wall is given next.

Start point or [STyle/Group/WIdth/Height/OFfset/Flip/Justify/Match/Arc]: **H**
Height <3000>: *Specify the desired height.*
Start point or [STyle/Group/WIdth/Height/OFfset/Flip/Justify/Match/Arc]: [Enter]

## OFfset

This option is used to set the distance of centerline of the wall from the wall edge. By default the value is set to zero. Figure 3-5 shows a wall being created using the **OFfset** option.

*Figure 3-5   The wall being created with centerline at its default position*

## Flip

This option is used to change the direction of wall to either side of the cursor. You can also use the CTRL key to change the direction of the wall.

## Justify

This option is used to switch the cursor position to the right, center, and left points of the wall. When you choose this option, four options will be available at the command prompt to justify the wall. The command prompt for justifying the wall after choosing the **Wall** tool is given next.

Start point or [STyle/Group/WIdth/Height/OFfset/Flip/Justify/Match/Arc]: J
Justification [Left/Center/Right/Baseline] <Baseline>: *Enter the justification option.*
Start point or [STyle/Group/WIdth/Height/OFfset/Flip/Justify/Match/Arc]: [Enter]

The justification options available at the prompt are discussed next.

### Left

The **Left** option is used to justify the wall on the left of the cursor.

### Center

The **Center** option is used to justify the wall on the center of the cursor.

### Right

The **Right** option is used to justify the wall on the right of the cursor.

### Baseline

The **Baseline** option is used to justify the wall on the center of the baseline.

You can also use SHIFT to toggle between the justification options.

## Match

This option is used to match the properties of the wall to be created with an existing wall. On choosing this option; you will be prompted to select a wall with which the properties will be matched. At the next prompt, enter the Property names to be matched. The options available at this prompt are Style, Group, Width, Height, Justify, and All. After specifying the options, you will be prompted to specify the start point of the wall. Specify the start point and end point of the wall to create a wall having properties similar to the one selected for matching.

## Arc

This option is used to create a wall in the shape of an arc. The command prompt for creating an arc shaped wall is given next.

Command: **WALLADD** ↵
Start point or [STyle/Group/WIdth/Height/OFfset/Flip/Justify/Match/Arc]: **Arc** ↵
Start point or [STyle/Group/WIdth/Height/OFfset/Flip/Justify/Match/Line]: *Specify the Start point of the wall segment.*
Mid point or [STyle/Group/WIdth/Height/OFfset/Flip/Justify/Match/Line]: *Specify the Mid point of the wall segment.*
End point or [STyle/Group/WIdth/Height/OFfset/Flip/Justify/Match/Line]: *Specify the End point of the wall segment.*
Mid point or
[STyle/Group/WIdth/Height/OFfset/Flip/Justify/Match/Line/Undo/Close]: *Specify the Mid point of the next wall segment or enter* **Close** *to close the wall segment.*
End point or
[STyle/Group/WIdth/Height/OFfset/Flip/Justify/Match/Line/Undo/Close]: *Specify the End point of the next wall segment or enter* **Close** *to close the wall segment.*

## Line

This option is used to create straight walls. This option will be available on the command prompt only when the **Arc** option is active. The command prompt for creating an arc shaped wall is already discussed.

## Undo

This option is used to invert the changes made by the previous command. Using the undo option, you can invert the changes made by the wall command in the current session.

## Close

This option is used to create a wall that makes a closed boundary by joining the other walls. When you choose this option, a wall connecting the first and last walls is created.

## ORtho close

This option is used to create two joining wall segments to make a closed boundary of walls. The command prompt for creating a wall using this option is given next.

Command: **WALLADD**
Start point or [STyle/Group/WIdth/Height/OFfset/Flip/Justify/Match/Arc]: *Specify the Start point of the wall segment.*
End point or [STyle/Group/WIdth/Height/OFfset/Flip/Justify/Match/Arc]: *Specify the End point of the wall segment.*
End point or [STyle/Group/WIdth/Height/OFfset/Flip/Justify/Match/Arc/Undo]: *Specify the End point of the wall segment.*
End point or
[STyle/Group/WIdth/Height/OFfset/Flip/Justify/Match/Arc/Undo/Close/ORtho close]: ORtho ⏎
Point on wall in direction of close: *Click on the direction in which you want the two perpendicular connecting walls to meet; the joining walls will be created, refer to Figure 3-6 and Figure 3-7.*

*Figure 3-6* *The corner point to be selected for the perpendicular walls*

*Figure 3-7* *The wall created after selecting the corner point using the **ORtho close** option*

You can also change the properties of a wall by using the **Properties** palette. Select the wall and then enter the **PR** command at the command prompt; the **Properties** palette will be displayed, as shown in Figure 3-8. Various rollouts available in this palette are discussed next.

## BASIC Rollout

The options available in this rollout are used for basic settings. There are three sub-rollouts available in this rollout which are discussed next.

### General Sub-rollout

The options in this sub-rollout are used for general settings such as the layer, description of the wall, and the segment type. You can also change the cleanup properties for the intersecting walls by using the options in this sub-rollout.

### Dimensions Sub-rollout

The options in this sub-rollout are used to specify the height, width, and length of the wall. You can also justify the wall by using the **Justify** option available in the sub-rollout.

### Location Sub-rollout

The options available in this sub-rollout are used to change the rotation angle and elevation of the wall. You can also change the insertion point and the orientation of wall by using the **Location** dialog box which is displayed on choosing the **Additional information** option available in this sub-rollout.

## ADVANCED Rollout

The options available in this rollout are used for changing the advanced settings such as cleanups, styles, and worksheets. These options are discussed next.

### Cleanups Sub-rollout

The options in this sub-rollout are used for changing the cleanup radius for intersecting walls. You can also apply overrides for the start and end cleanup radius by using the options available in this sub-rollout.

### Style overrides Sub-rollout

The options in this sub-rollout are used to override

*Figure 3-8  The **Properties** palette displayed on selecting a wall*

the starting and ending endcap styles. You can also set the priority of a wall with respect to an intersecting wall by using the **Priority Overrides** option available in this sub-rollout.

### Worksheets Sub-rollout

The options in this sub-rollout are used to modify the cross-section of the wall. You can also specify the start point and the end point of the modifiers by using the options in this rollout.

## Curtain Wall

This tool is used to create a curtain wall. A curtain wall is a non-structural wall which is used to avoid the effect of weather on the building. This wall does not support any load in the building. The curtain walls created by using the **Curtain Wall** tool can be straight or curved.

The command prompt for creating a curtain wall is given next.

Command: CURTAINWALLADD
Start point or [STyle/Height/Match/Arc]: *Specify the Start point of the wall segment.*
End point or [STyle/Height/Match/Arc]: *Specify the end point of the current wall segment and start point of the next wall segment.*
End point or [STyle/Height/Match/Arc/Undo]: *Specify the end point of the current wall segment and start point of the next wall segment.*
End point or [STyle/Height/Match/Arc/Undo/Close/ORtho close]: *Specify the end point of the current wall segment and start point of the next wall segment or enter **Close** to create a closing wall.*

The options available at the command prompt are already discussed. You can also use the **Properties** palette to change properties of the created wall. The options available in the **Properties** palette for a curtain wall are discussed next.

## BASIC Rollout
The options available in this rollout are used for basic settings. Most of the options available in this rollout are same as discussed earlier. Rest of the options are discussed next.

### Dimensions Sub-rollout
The options in this sub-rollout are used to change the base height, length, start miter, and end miter.

## ADVANCED Rollout
The options available in this rollout are used to change the advanced settings. The options available in this rollout are already discussed.

## Curtain Wall Unit
 This tool is used to create a unit of curtain walls. On choosing this tool, you will be prompted to specify the start point of the wall. Specify the start point of the wall; you will be prompted to specify the end point of the wall. Specify the end point of the wall; you will be prompted to specify the height of the wall. On specifying the height, the **Curtain Wall Unit Styles** dialog box will be displayed, as shown in Figure 3-9.

*Figure 3-9  The **Curtain Wall Unit Styles** dialog box*

Select a curtain wall unit style from the dialog box and choose the **OK** button; the wall unit with the defined specifications will be created in the drawing area. You can change the specifications and properties of the wall unit by using the **Properties** palette.

Using the **Properties** palette, you can change the properties of a curtain wall unit. The options available for a curtain wall unit are same as discussed earlier.

# CREATING DOORS

Doors are the movable structures used to close the entrance of a room or building. The tools to create a door are available in the **Door** drop-down of the **Build** panel in the **Home** tab of the **Ribbon**, refer to Figure 3-10. There are three tools available in the **Door** drop-down: **Door**, **Opening**, and **Door/Window Assembly**. These tools are discussed next.

*Figure 3-10  The **Door** drop-down*

## Door

This tool is used to create doors of a specified profile. To create a door, choose the **Door** tool from the **Door** drop-down; the **Properties** palette will be displayed, refer to Figure 3-11 and you will be prompted to select a wall or a grid assembly. You can press ENTER to create a door as an individual entity or you can attach the door with a wall. To attach a door to the wall, select an existing wall; you will be prompted to select an insertion point. Select the desired point on the wall; the door will be created with the parameters specified in the **Properties** palette. The options available in the **Dimensions** rollout of the palette are discussed next.

*Figure 3-11  The **Properties** palette displayed on choosing the **Door** tool*

## Standard sizes

This drop-down list displays various standard sizes created by using the **Style Manager**. You can choose any of the standard sizes for the door or you can specify a custom size by using the options available in the **Properties** palette.

**Tip**
*To add an arched door, choose the **Style Manager** tool from the **Style & Display** panel in the **Manage** tab; the **Style Manager** will be displayed. In the **Style Manager**, create a new door style with arch shape. The **Style Manager** will be discussed in later chapters.*

## Rise

This option will be available only for an arched door or a door pointed at the top. This option is used to specify the distance of arch or top corner point from the top endpoint of the door supports, refer to Figure 3-12.

*Figure 3-12  Specifications of an arched door*

## Width

This option is used to specify the width of the door, refer to Figure 3-12. This option is used to customize a standard door. By default, the width specified by Door style is displayed in this field. You can change the value in this field as per your requirement.

## Height

This option is used to specify the height of the door, refer to Figure 3-12. This option is used to customize a standard door. By default, the height specified in the Door style is displayed in this field. You can change the value in this field as per your requirement.

## Measure to

The options available in this drop-down are used to specify whether the width and height of the door is measured from the inner edge of the frame or the outer edge of the frame.

## Swing angle

This option is used to specify the maximum angle at which the door can be opened.

You can change the parameters related to location of door by using the options available in the **Location** rollout. Some of the options available in the **Location** rollout of the palette are discussed next.

## Vertical alignment

The options in this drop-down list will be available only if the selected door is anchored. Using these options, you can specify whether the height of door will be measured from the threshold of the door or the head of the door.

## Head Height / Threshold Height

The **Head Height** or **Threshold Height** option gets activated depending on the selection made in the **Vertical alignment** drop-down list. Using these options, you can specify the height of door from the head/threshold.

Rest of the options in the **Properties** palette have already been discussed.

# Opening

This tool is used to create an opening of a specified profile in the wall. To create an opening, choose the **Opening** tool from the **Door** drop-down in the **Build** panel of the **Home** tab; you will be prompted to select a wall or press ENTER. Select the wall on which you want to create an opening; you will be prompted to specify the insertion point. Click to specify the insertion point on the wall; the opening will be created. On selecting an opening, the **Properties** palette is displayed, as shown in Figure 3-13. Some of the options available in the **Properties** palette that are used to change the properties of an opening are discussed next.

## Shape

This drop-down list is available in the **General** sub-rollout of the **Basic** rollout. The options in this drop-down are used to change the shape of the opening. There are 13 types of predefined shapes available in this drop-down for creating an opening. You can customize the shape of an opening by selecting the **Custom** option from the drop-down list. On selecting the **Custom** option, the **Profile** drop-down list will be displayed below the **Shape** drop-down list. You can create the selected custom profiles of opening using this drop-down list.

## Width

This option is available in the **Dimensions** sub-rollout. Using this option, you can specify the width of opening.

## Height

This option is available in the **Dimensions** sub-rollout. Using this option, you can specify the height of opening.

**Figure 3-13** *The* **Properties** *palette displayed on selecting an opening*

## Vertical alignment

This drop-down list is available in the **Location** sub-rollout. The options in this drop-down will be available only if the selected opening is anchored. Using these options, you can specify whether the height of opening will be measured from the threshold of the opening or the head of the opening.

## Door/Window Assembly

This tool is used to create a door or window assembly. The assembly contains required number of doors, windows, or both. To create a door/window assembly, choose the **Door/Window Assembly** tool from the **Door** drop-down available in the **Build** panel of the **Home** tab; you will be prompted to select a wall, a grid assembly, or press ENTER. Select a wall; you will be prompted to specify the insertion point for door/window assembly. Click at the desired point on the wall; an anchored door/window assembly will be created. The options available in the **Properties** palette for a door/window assembly are same as discussed earlier.

# CREATING WINDOW

Windows are openings in a wall or door to facilitate the passage to light. If not closed, they also allow air and sound to pass through the wall or door. In AutoCAD MEP, windows are created by using the tools available in the **Window** drop-down. This drop-down is available in the **Build** panel of the **Home** tab. The options available in this drop-down are discussed next.

# Window

This tool is used to create a standard window. To create a window, choose the **Window** tool from the **Window** drop-down available in the **Build** panel; you will be prompted to select a wall, a grid assembly, or press ENTER. Select a wall; you will be prompted to specify an insertion point. Click on the wall to specify the point; the window will be created at the specified point. Click on the wall to create more windows or press ENTER to exit the command. You can set the opening percentage of window by using the **Opening percent** field available in the **Properties** palette. The other options available in the **Properties** palette have already been discussed.

# Corner Window

This tool is used to create a window at the position where two walls meet. This window can only be created on a cornered wall. To create a corner window, choose the **Corner Window** tool from the **Window** drop-down available in the **Build** panel; you will be prompted to select a wall. Select a wall which is connected to the other wall at the corner; you will be prompted to specify the insertion point. Click on the wall to place the window and press ENTER to exit the command.

There are three conditions for creating a corner window;
1. The walls meeting to create a corner must be in "L" shape.
2. Wall joints must be created by linear walls only.
3. Wall joints must be cleaned properly.

Various options available in the **Dimensions** sub-rollout of the **Properties** palette are discussed next.

### Standard sizes
The options in this drop-down list are used to change the standard size of the corner window. The standard sizes are generated by using the **Style Manager**. The pattern of size displayed in this drop-down list is given next.

(**Width 1** x **Width 2**) x **Height**

Here **Width 1** and **Width 2** are width of the window in two different directions, and **Height** is the height of window from bottom edge to top edge.

### Width
This field shows the total width of window. This is the sum of Width 1 value and Width 2 value. You cannot change the value of this field.

### Width 1
This edit box is used to change the value of width in the first direction. When you click on this edit box, a button will be displayed on the right of the edit box. On choosing this button, you can specify the value of width by selecting two points in the drawing. The value changed in this edit box is also reflected in the **Width** field and the **Standard sizes** drop-down list.

## Width 2
This edit box is used to change the value of width in the second direction. When you click on this edit box, a button will be displayed on the right of the edit box. On choosing this button, you can specify the value of width by selecting two points in the drawing. The value changed in this edit box is also reflected in the **Width** field and the **Standard sizes** drop-down list.

## Height
This edit box is used to change the value of height of the window.

## Measure to
The options available in this drop-down are used to specify whether the value of width and height will be measured from inside of the frame or outside of the frame.

## Opening measure
The options available in this drop-down are used to specify the width measurement pattern. There are three options available in this drop-down: **Inside of opening**, **Outside of opening**, and **Center of opening**.

### Inside of opening
When you choose this option, the width of the first side of corner window will be measured from the inner edge of the other side of the corner window.

### Outside of opening
When you choose this option, the width of the first side of corner window will be measured from the outer edge of the other side of the corner window.

### Center of opening
When you choose this option, the width of first side of corner window will be measured from the center of the edge of other side of the corner window.

## Opening percent
Using this edit box, you can specify the opening percentage of the window.

# CREATING ROOFS AND SLABS
The tools to create roofs and slabs are available in the **Roof Slab** drop-down, refer to Figure 3-14. The tools in this drop-down are **Roof Slab**, **Roof**, and **Slab**. These tools are discussed next.

## Roof Slab

This tool is used to create a segment of roof. This segment does not have a direct connection with other entities. To create a roof slab, choose the **Roof Slab** tool from the **Roof Slab** drop-down available in the **Build** panel of the **Home** tab; you will be prompted to specify the start point of the roof slab. The command sequence for creating a roof slab is given next.

*Figure 3-14  The Roof Slab drop-down*

Specify start point or [STyle/MOde/Height/Thickness/SLope/OVerhang/Justify/MAtch]: *Specify the Start point of the roof slab.*

Specify next point or [STyle/MOde/Height/Thickness/SLope/OVerhang/Justify/MAtch]: *Specify the next point of the wall segment.*

Specify next point or [STyle/MOde/Height/Thickness/SLope/OVerhang/Justify/MAtch/Undo/Ortho close]: *Specify the next point of the wall segment.*

Specify next point or [STyle/MOde/Height/Thickness/SLope/OVerhang/Justify/MAtch/Undo/Ortho close/Close]: *Specify the next point of the wall segment or ENTER Close to close the boundary.*

Specify next point or [STyle/MOde/Height/Thickness/SLope/OVerhang/Justify/MAtch/Undo/Ortho close/Close]: *Specify the next point of the wall segment or ENTER **Close** to close the boundary.*

After selecting a roof slab, the **Properties** palette will be displayed, as shown in Figure 3-15. Various options available in the **Dimensions** rollout of this palette are discussed next.

*Figure 3-15  The **Properties** palette displayed on selecting a roof slab*

## Thickness

This edit box is used to specify the thickness of the roof slab. When you click in this edit box, a button is displayed on the right of this edit box. On choosing this button, you can specify the value of thickness by selecting two points.

## Vertical offset

This edit box is used to specify the offset value of slab from the original position in vertical direction. You can also specify the offset value by selecting two points.

## Horizontal offset

This edit box is used to specify the offset value of slab from the original position in horizontal direction. You can also specify the offset value by selecting two points.

## Edges

This button is available below the **Horizontal offset** edit box. When you choose this button, the **Roof Slab Edges** dialog box will be displayed, as shown in Figure 3-16. Also, the preview of the changes made in the edges are displayed in the **Preview** area on the right in the dialog box. You can change the parameters such as overhang value, edge style, and edge cut type by using the table displayed on the left in the dialog box. Various columns available in this table are discussed next.

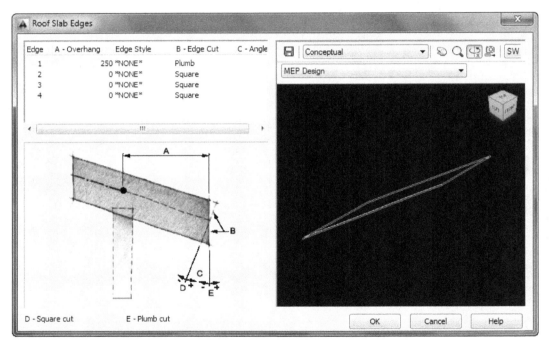

*Figure 3-16   The **Roof Slab Edges** dialog box*

## Edge

All edges of the selected roof slab are displayed in this column by their sequence numbers. When you select any sequence number, its respective edge is highlighted in the **Preview** area.

## A-Overhang

Using the fields available in this column, you can change the overhang value. Overhang is the area of slab which is out of the support boundary.

### Edge Style

By default, the values in this column are set to **NONE**. You can choose any predefined style for slab edges. You can define an **Edge Style** by using the **Roof Slab Edge Style** node in the **Style Manager**.

### B-Edge Cut

In this column, edge cut methods are displayed. There are two options for edge cut: **Square** and **Plumb**. The **Square** option is used to cut the edge perpendicular to the roof slab and the **Plumb** option is used to cut the edge perpendicular to the ground.

### C-Angle

In this column, you can specify an angle for the edge. If you enter a positive value, the bottom edge will extend. If a negative value is entered, then the top edge will extend.

## Rise

This edit box is used to specify the value of rise for a roof slab. Rise is the slope difference between the start and end edges of the slope of roof slab along vertical direction. You can also specify the value of rise by selecting two points. To do so, choose the button displayed at the right of this edit box; you will be prompted to select points. Select the points to specify the rise value.

## Run

This field is used to display the value of run for a roof slab. Run is the horizontal span of the slope. The value in this field is generally fixed.

## Slope

This edit box is used to specify the value of slope for a roof slab. The value of slope is specified in degrees. Any change in the value specified in the **Rise** edit box or the **Run** edit box is reflected in the **Slope** edit box.

## Hold fascia elevation

The options in this drop-down are used to manage the fascia of roof. There are three options available in this drop-down: **No**, **By adjusting overhang**, and **By adjusting baseline height**.

### No

When you choose this option, the alignment of fascia is ignored.

### By adjusting overhang

When you choose this option, the fascia is aligned with the roof slab by adjusting the overhang.

### By adjusting baseline height

When you choose this option, the fascia is aligned with the roof slab by adjusting the baseline height.

## Pivot Point X

This edit box is used to specify the position of pivot point of roof slab in X direction. You can calculate the value of position in X direction by using the **QuickCalc** button available on the right of the edit box. You can also use the **Pick Point** tool to specify the pivot point location which is discussed in the next section.

## Pivot Point Y

This edit box is used to specify the position of pivot point of roof slab in Y direction. Alternatively, you can specify the position of point by using the **Pick Point** tool available on the right of the edit box. You can also use the **QuickCalc** button as discussed earlier.

## Pivot Point Z

This edit box is used to specify the position of pivot point of roof slab in Z direction. Alternatively, you can specify the position of point by using the **Pick Point** tool available on the right of the edit box. You can also use the **QuickCalc** button as discussed earlier.

# Roof

This tool is used to create a multi-peaked roof. To create a multi-peaked roof, invoke the **Roof** tool from the **Roof Slab** drop-down available in the **Build** panel of the **Home** tab; you will be prompted to specify the roof points. Specify the roof points; the roof will be created with the specified roof points. The **Properties** palette on selecting a roof will be displayed, as shown in Figure 3-17.

*Figure 3-17  The **Properties** palette displayed on selecting a roof*

The options available in the **Lower Slope** rollout of the **Properties** palette are discussed next. The options available in the **General**, **Dimensions**, and **Location** rollouts are same as discussed earlier. The options available in the **Lower Slope** rollout are discussed next.

## Plate height

This edit box is used to specify the height of the base point of roof from the ground. Alternatively, you can specify the plate height by using two points. In this case, you need to choose the **Pick Point** button available on the right of the edit box.

## Rise

This edit box is used to specify the value of rise for a roof. Rise is the difference in start and end edges of the slope of roof along vertical direction. You can also specify the value of rise by selecting two points. To do so, choose the button displayed at the right of this edit box; you will be prompted to select points. Select any two points on the screen to specify the rise value.

## Run

This field is used to display the value of run for a roof. Run is the horizontal span of the slope. The value in this field is generally fixed.

## Slope

This edit box is used to specify the value of slope for a roof. The value of slope is specified in degrees. Any change in the value specified in the **Rise** edit box or the **Run** edit box is reflected in the **Slope** edit box.

You can also specify the plate height, rise, and slope value for the upper slope of the roof. To specify the values for upper slope, you need to select the **Double slope** option from the **Shape** drop-down in the **Dimensions** rollout. On doing so, the **Upper Slope** rollout will be displayed. The options in this rollout are similar to the options discussed for **Lower Slope** rollout.

# Slab

This tool is used to create flat roofs or floors. The roofs or floors created by using this tool are in the form of a segment. This segment does not have a direct connection with other entities. To create a slab, choose the **Slab** tool from the **Roof Slab** drop-down available in the **Build** panel of the **Home** tab; you will be prompted to specify the start point of the slab. The command prompt for creating a slab is given next.

Specify start point or [STyle/MOde/Height/Thickness/SLope/OVerhang/Justify/MAtch]: *Specify the Start point of the roof slab.*
Specify next point or [STyle/MOde/Height/Thickness/SLope/OVerhang/Justify/MAtch]: *Specify the next point of the wall segment.*
Specify next point or [STyle/MOde/Height/Thickness/SLope/OVerhang/Justify/MAtch/Undo/Ortho close]: *Specify the next point of the wall segment.*
Specify next point or [STyle/MOde/Height/Thickness/SLope/OVerhang/Justify/MAtch/Undo/Ortho close/Close]: *Specify the next point of the wall segment.*
Specify next point or [STyle/MOde/Height/Thickness/SLope/OVerhang/Justify/MAtch/Undo/Ortho close/Close]: **C**
Specify start point or
[STyle/MOde/Height/Thickness/SLope/OVerhang/Justify/MAtch/Undo]:⏎

After selecting a slab, the **Properties** palette is displayed, as shown in Figure 3-18. The options in the palette have already been discussed.

*Figure 3-18  The **Properties** palette displayed on selecting a slab*

# CREATING STAIRS AND RAILINGS

Stairs are the structures that enable transportation from one floor to the other. Railings are the structures which act as support or barrier on the stairs. The options to create stairs and railings are available in the **Stair** drop-down. These options are discussed next.

## Stair

This tool is used to create rectangular stairs. There are various shapes of stairs such as straight, multi-landing, spiral, L-shaped, and U-shaped. While going on a curved path, the stairs can have curved landing. In such cases, the edges of stairs can also be curved. To create a stair, choose the **Stair** tool from the **Stair** drop-down available in the **Build** panel of the **Home** tab; you will be prompted to specify the start point of the stairs. Specify a point as the start point of the stairs; you will be prompted to specify the end point of the stairs. Specify a point as the end point to complete the stairs. While specifying the end point, you can specify the landing at desired position but this position must be within the specified limit. Limit of a stair is counted by the value specified in the **Riser count** edit box. The **Properties** palette after choosing the **Stair** tool is displayed, as shown in Figure 3-19. Some important options available in the **General** rollout of the palette on choosing this tool are discussed next.

## Shape

The options available in this drop-down are used to specify the shape of the stairs. There are four options available in this drop-down; **U-shaped**, **Multi-landing**, **Spiral**, and **Straight**. These options are discussed next.

### U-shaped

Using this option, you can create U-shaped stairs, refer to Figure 3-20. In this type of stairs, the two rows of stairs run parallel and meet at the end points by a half-landing.

### Multi-landing

Using this option, you can create stairs having multiple landings, refer to Figure 3-20. You can provide quarter landing or half landing. This landing can be flat or can have a turn.

### Spiral

Using this option, you can create stairs in the shape of spiral, refer to Figure 3-20. In this type of stairs, the steps of stair revolve about a common center point from bottom to top. Spiral stairs can be in clockwise direction or counter-clockwise direction.

*Figure 3-19  The **Properties** palette displayed on choosing the **Stair** tool*

### Straight

Using this option, you can create straight stairs, refer to Figure 3-20. In this type of stairs, the steps are created on a straight path that is inclined to the horizontal plane.

## Turn type

This drop-down is displayed only when **U-shaped** or **Multi-landing** is selected from the **Shape** drop-down. The options available in this drop-down are used to define the type of landing or turning for the U-shaped or multi-landing stairs. There are four options available in this drop-down that are discussed next.

### 1/2 landing

Using this option, you can create U-shaped or multi-landing stairs having a flat landing of user-defined length, refer to Figure 3-21.

### 1/2 turn

Using this option, you can create U-shaped or multi-landing stairs having steps at the turn, refer to Figure 3-21. Note that there must be at least three segments in the stairs and direction of turn of all the segments must be the same.

## 1/4 landing

This option is available only when **Multi-landing** is selected from the **Shape** drop-down. When you select this option, the width of landing created at turn will be equal to the width of the stair, refer to Figure 3-21. Generally, these type of stairs have two or more steps joined to the square landing.

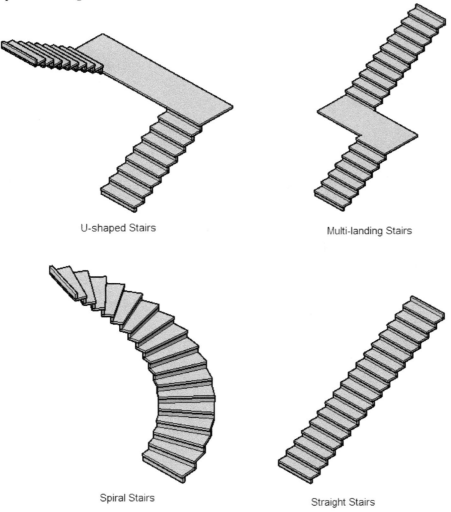

U-shaped Stairs                                      Multi-landing Stairs

Spiral Stairs                                        Straight Stairs

*Figure 3-20  Various type of stairs available in AutoCAD MEP*

## 1/4 turn

This option is available only when **Multi-landing** is selected from the **Shape** drop-down. When you select this option, the resulting stairs will have steps at the turn, refer to Figure 3-21. The stairs can turn in both directions.

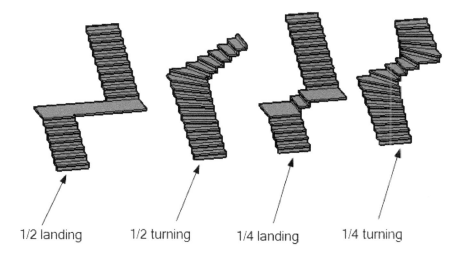

1/2 landing      1/2 turning      1/4 landing      1/4 turning

*Figure 3-21*  *Stairs with different turning types selected*

### Winder Style

This drop-down is available only when **1/2 turn** or **1/4 turn** is selected from the **Turn type** drop-down. Various winder styles are available in this drop-down. By default, the **Balanced** style is selected from this drop-down. There are three types of winder styles available in AutoCAD MEP: **Balanced**, **Single Point**, and **Manual**. The **Balanced** winder style is used to distribute steps of stairs evenly throughout the run. The **Single Point** winder style is used to align the edges of steps or riser faces with the winder turn centre. The **Manual** winder style can be based on any of the two winder styles discussed earlier. In this type of winder style, you can manually assign riser condition to an individual step.

### Horizontal Orientation

This drop-down is displayed only when **U-shaped** or **Spiral** is selected from the **Shape** drop-down. The options available in this drop-down are used to define the direction of turn of stairs. There are two options available in this drop-down: **Clockwise** and **Counterclockwise**. Using these options, you can select the clockwise or counterclockwise direction of turn for the stairs.

### Vertical Orientation

The options in this drop-down are used to specify the vertical direction of the stairs. There are two options available in this drop-down, **Up** and **Down**. The **Up** option is selected when you need to create stairs from lower floor to higher floor. The **Down** option is selected when you need to create stairs from higher floor to lower floor.

The options in the **General** rollout are also used to control the shape of stairs. Some important terms for defining the size of stairs is displayed in Figure 3-22. The options available in the **Dimensions** rollout of the palette are used to control the size of stairs. These options are discussed next.

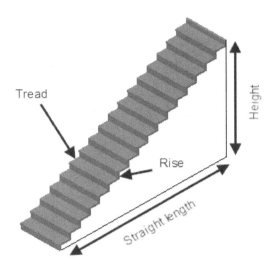

*Figure 3-22* Some important terms for defining the size of stairs

## Width

This edit box is used to specify the width of a step. You can also specify the width of a step by using two points. To do so, click on the button displayed on the right of the edit box; you will be prompted to select start point for width calculation. Select a start point; you will be prompted to select an end point for width calculation. Select a point; the width value for the specified points will be reflected in the edit box.

## Height

This edit box is used to specify total height of the stairs. You can also specify the total height of stairs by using two points. The method of using the points is already discussed.

## Justify

The options in this drop-down are used to change the justification of stairs. There are three options available in this drop-down: **Left**, **Right**, and **Center**. These options are used to justify the stairs in left, right, and center, respectively.

## Terminate with

The options in this drop-down are used to change the justification of stairs. There are three options available in this drop-down: **Riser**, **Tread**, and **Landing**. When you select the **Riser** option, the stairs end with a riser. When you select the **Tread** option, the stairs end with a tread. If the **Landing** option is selected, the stairs end with a landing. Figure 3-23 shows the output of various options selected in the **Terminate with** drop-down.

*Figure 3-23   The output of various options selected in the **Terminate with** drop-down*

## Flight length

The options in this drop-down are used to control the length of one segment of a staircase. There are two options available in this drop-down: **Distance** and **Tread length**. Using the **Distance** option, you can specify the flight length by specifying total distance value. Using the **Tread length** option, you can specify the flight length by specifying length of one tread. The total flight length is calculated on the basis of number of treads in the staircase.

## Calculation rules

When you click on the **Tread** button displayed in this field, the **Calculation Rules** dialog box will be displayed, as shown in Figure 3-24. By default, the **D-Tread** edit box is activated. On changing the value in this edit box, the value in the other edit boxes in this dialog box will change accordingly. To change values in other edit boxes manually, you need to lock the **D -Tread** edit box. To do so, choose the **lock** button; the **Calculation Rules** dialog box will be displayed, as shown in Figure 3-25. Now, you can change the value of any of the edit box in the dialog box by choosing the button adjacent to the respective edit box. In this dialog box, you can manually define the value in two edit boxes simultaneously. But if you choose the **C-Riser** edit box for defining the value of riser manually, then the other three edit boxes will be locked.

*Figure 3-24  The **Calculation Rules** dialog box*

The value of the edit boxes that are set to change manually can also be changed directly using the **Properties** palette. Note that in the **Properties** palette only those edit boxes will be active that are set to change manually in the **Calculation Rules** dialog box.

*Figure 3-25  The **Calculation Rules** dialog box displayed after locking the edit boxes*

Some of the options in the **Properties** palette are displayed only when **Spiral** is selected from the **Shape** drop-down of the palette. These options are discussed next.

## Specify on screen

The options in this drop-down are used to specify whether the second point specifies the radius along the starting direction or not. There are two options available in this drop-down: **Yes** and **No**. When you choose **Yes** from this drop-down, then you can specify the radius of stairs on the screen while creating the stairs.

## Radius

This edit box is available only when you select the **No** option from the **Specify on screen** drop-down. Using this edit box, you can specify the value of radius for the spiral stairs.

## Arc constraint

The options in this drop-down are used to constrain the arc of the spiral stairs. There are three options available in this drop-down: **Free, Total degrees**, and **Degrees per tread**. When the **Free** option is selected from this drop-down, the stair run is unconstrained. When **Total degrees** is selected from this drop-down, you can specify the total angle of arc around which the stairs are created. When you select the **Degrees per tread** option, you can specify the angle of each tread with respect to the adjacent in the stairs.

## Arc angle

This edit box is active only when the **Total degrees** or the **Degrees par tread** option is selected from the **Arc constraint** drop-down. Using this edit box, you can specify the value of arc angle. The value specified in this edit box is linked to the option selected in the **Arc constraint** drop-down. If you have selected the **Total degrees** option from the **Arc constraint** drop-down then the value specified in the **Arc angle** edit box is applicable for total degree of the spiral stairs. If the **Degrees per tread** is selected from the **Arc constraint** drop-down, the angle value specified in the **Arc angle** edit box is applicable for each tread.

Other options in the **Dimensions** rollout have already been discussed. Some of the important options available in the **Advanced** rollout of the **Properties** palette are discussed next.

## Top offset

This edit box is available in the **Floor Settings** rollout. This edit box is used to specify the value for thickness of the floor at the top of the stairs.

## Bottom offset

This edit box is also available in the **Floor Settings** rollout and is used to specify the value of thickness of the floor at the bottom of the stairs.

## Minimum limit type

Using the options available in this drop-down, you can change the minimum limit for the height of flight. There are three options available in this drop-down: **NONE, Risers**, and **Height**. By default, the **NONE** option is selected in this drop-down. As a result, no limit is set for height or risers of the stairs. You can specify the minimum limit of risers by selecting the **Risers** option from the drop-down. You can also specify the minimum height of stairs by using the **Height** option from this drop-down.

## Minimum Risers

This edit box is available in the **Flight Height** rollout only when the **Risers** option is selected from the **Minimum limit type** drop-down. You can specify the minimum value for height of the riser by using this edit box.

## Minimum Height

This edit box is available in the **Flight Height** rollout only when the **Height** option is selected in the **Minimum limit type** drop-down. You can specify the minimum value for height of the stairs by using this edit box.

## Maximum limit type

Using the options available in this drop-down, you can change the maximum limit for the height of flight. There are three options available in this drop-down: **NONE**, **Risers**, and **Height**. By default, the **NONE** option is selected in this drop-down. As a result, no limit is set for height or risers of the stairs. You can specify the maximum limit of risers by selecting the **Risers** option from the drop-down. You can also specify the maximum height of stairs by using the **Height** option from this drop-down.

## Maximum Risers

This edit box is available in the **Flight Height** rollout only when the **Risers** option is selected in the **Maximum limit type** drop-down. You can specify the maximum value for height of the riser by using this edit box.

## Maximum Height

This edit box is available in the **Flight Height** rollout only when the **Height** option is selected in the **Maximum limit type** drop-down. You can specify the maximum value for height for the stairs by using this edit box.

## Headroom height

This edit box is available in the **Interference** rollout. This edit box is used to specify the height of ceiling around the stairs opening from the tread. This height is required so that people do not collide with the roof while going through the stairs.

## Left clearance

This edit box will be available in the **Interference** rollout only when **Straight** or **Multi-landing** is selected in the **Shape** drop-down. This edit box is used to specify the clearance value for stairs from the left edge of the opening in the roof.

## Right clearance

This edit box will be available in the **Interference** rollout only when **Straight** or **Multi-landing** is selected in the **Shape** drop-down. This edit box is used to specify the clearance value for stairs from the right edge of the opening in the roof.

## Inside clearance

This edit box will be available in the **Interference** rollout only when **Spiral** or **U-shaped** is selected in the **Shape** drop-down. This edit box is used to specify the clearance value for stairs from the inner edge of the opening in the roof.

## Outside clearance

This edit box will be available in the **Interference** rollout only when **Spiral** or **U-shaped** is selected in the **Shape** drop-down. This edit box is used to specify the clearance value for stairs from the outer edge of the opening in the roof.

## Component

This option is available in the **Worksheet** rollout. It is used to edit the parameters for stairs. When you choose this option, the **Stair Components** dialog box is displayed, as shown in Figure 3-26. The options available in this dialog box are discussed next.

### Tread

This check box is available in the **Flight Dimensions** area. This check box is used to enable change in the value of tread thickness.

### Riser

This check box is available in the **Flight Dimensions** area. This check box is used to enable change in the value of riser thickness.

### A - Tread Thickness

This edit box is active only when the **Tread** check box is selected. Using this edit box, you can specify the value for tread thickness. An annotated preview of all the related parameters is displayed in the left area.

*Figure 3-26 The **Stair Components** dialog box*

### B - Riser Thickness

This edit box is active only when the **Riser** check box is selected. Using this edit box, you can specify the value for riser thickness. An annotated preview of all the related parameters is displayed in the left area.

### C - Nosing Length

This edit box is used to specify the length for extended portion of tread over the riser.

### Sloping Riser

Select this check box if you want to create riser with a slope.

### D - Landing Thickness
This edit box is available in the **Landing Dimensions** area. This edit box is used to specify the value for thickness of the landing.

### E - Additional Width
This edit box is available in the **Landing Dimensions** area. This edit box is used to specify the value for additional width of landing.

### Reset to Style Values
This option is used to reset values of all the parameters to their original values. This button is available at the bottom of the dialog box.

## Landing extensions
This tool is available in the **Worksheet** rollout. This tool is used to edit the parameters for stairs related to landing extensions. When you choose this tool, the **Landing Extensions** dialog box will be displayed, as shown in Figure 3-27. The options available in this dialog box are discussed next.

*Figure 3-27  The **Landing Extensions** dialog box*

**Note**
*You can edit the parameters in this dialog box only when you have selected the **Allow Each Stair to Vary** check box from the **Landing Extensions** tab of the **Stair Styles** dialog box in the **Style Manager**. The **Style Manager** is displayed on choosing the **Style Manager** tool from the **Style & Display** panel in the **Manage** tab of the **Ribbon**.*

### A - Distance to First Tread DOWN

This edit box is used to specify the value of extension of the tread attached to the lower riser. By specifying this value, you can extend the landing up to desired distance towards lower riser.

### Add Tread Depth

Select this check box if you want to add tread depth to the value specified in the **Distance to First Tread DOWN** edit box.

### B - Distance to First Tread UP

This edit box is used to specify the value for extension of the tread attached to the higher riser. By specifying this value, you can extend the landing up to desired distance towards higher riser.

### Add Tread Depth

Select this check box if you want to add tread depth to the value specified in the **Distance to First Tread UP** edit box.

### Extend Landings to Prevent Risers and Treads Sitting under Landings

This check box is available below the preview area. Select this check box if you want to create landing at the level of the adjacent riser and tread. If you clear this check box, you will get flush or rectangular landings.

### Extend Landings to Merge Flight Stringers with Landing Stringers

This check box is available in the **Stringer Resolution** area of the dialog box. Select this check box to extend the landings so that the landing stringers merge with the flight stringer.

# Railing

This tool is available in the **Stair** drop-down of the **Build** panel in the **Home** tab. This tool is used to create a stand-alone railing as well as a railing attached to stairs or some objects. To create a railing, choose the **Railing** tool from the **Stair** drop-down; you will be prompted to specify the start point of the railing. Select a point as the start point; you will be prompted to specify the end point of the railing. Select a point as the end point; you will be prompted to specify the end point again. You can specify an end point for the next segment or you can press ENTER to exit the command. Figure 3-28 shows an annotated railing created with guardrail, handrail, and bottomrail. Figure 3-29 shows the **Properties** palette displayed on selecting a railing.

Some of the important options available in the **Dimension** rollout of the **Properties** palette are discussed next.

## Rail locations

When you choose this option, the **Rail Locations** dialog box will be displayed, as shown in Figure 3-30. Various options available in this dialog box are discussed next.

**Figure 3-28**  *An annotated railing*

**Figure 3-29**  *The **Properties** palette displayed on selecting a railing*

*Figure 3-30  The **Rail Locations** dialog box*

 **Note**
*The options in this dialog box can be edited only when the **Allow Each Railing to Vary** check box is selected from the **Rail Locations** tab of Railing Styles in the **Style Manager**.*

**Guardrail**

When you select this check box, the edit boxes available next to the check box get activated. Using these edit boxes, you can specify horizontal height, sloping height, and offset value of guardrail from the post. You can also specify the side of offset by using the options available in the **Side for Offset** drop-down.

**Handrail**

When you select this check box, the edit boxes available next to the check box get activated. Using these edit boxes, you can specify horizontal height, sloping height, and offset value of handrail from the post. You can also specify the side of offset by using the options available in the **Side for Offset** drop-down.

**Bottomrail**

When you select this check box, the edit boxes available next to the check box get activated. Using these edit boxes, you can specify horizontal height, sloping height, and offset value for bottomrail from the post. You can also specify the side of offset by using the options available in the **Side for Offset** drop-down.

When you select the **Bottomrail** check box, then the **Number of Rails** and **Spacing of Rails** edit boxes get enabled. Using the **Number of Rails** edit box, you can specify the number of rails to be created from the bottom. Using the **Spacing of Rails** edit box, you can specify the value of spacing between two successive bottom rails.

## Post locations

When you choose this tool, the **Post Locations** dialog box will be displayed, as shown in Figure 3-31. The options available in this dialog box are discussed next.

**Note**
*The options in this dialog box can be edited only when the **Allow Each Railing to Vary** check box is selected from the **Post Locations** tab of Railing Styles in the **Style Manager**.*

### Fixed Posts

This check box is used to enable the edit boxes that are used to modify the value of extension of the post. There are two edit boxes to change the value of extension: **Extension of ALL Posts from Top Railing** and **Extension of ALL Posts from Floor level**. The **Extension of All Posts from Top Railing** edit box is used to specify the value of post above or below the top railing. The **Extension of ALL Posts from Floor level** is used to specify the height of all posts from the floor.

*Figure 3-31  The **Post Locations** dialog box*

### Fixed Posts at Railing Corners

This check box is used to include the posts present at the railing corners.

### Dynamic Posts

When you select this check box, the **Maximum Center to Center Spacing** edit box is enabled. This edit box is used to modify the distance between two successive dynamic posts. Specify the value for distance between the dynamic posts in the edit box; the value will be taken as the maximum limit of distance between the two successive posts. When you create a railing having length more than this value, a new post is added in the railing.

### Balusters

When you select this check box, the **Extension of Balusters from Floor Level** and **Maximum Center to Center Spacing** edit boxes are enabled. These check boxes are used to modify the values related to baluster. The **Extension of Balusters from Floor Level** edit box is used to specify the distance for bottom point of baluster from the level of corresponding tread. The **Maximum Center to Center Spacing** edit box is used to specify the center to center distance between the consecutive balusters.

### Stair Tread Length Override

When you select this check box, the **Number per Tread** edit box is enabled. This edit box is used to override the value for number of balusters per tread.

### Perpendicular posts

There are two options available in this drop-down: **Yes** and **No**. When **Yes** option is selected from this drop-down, the posts in the railing will be perpendicular to the rails in the railing. If you select the **No** option from this drop-down, the posts will be perpendicular to the tread.

## Stair Tower

This tool is available in the **Stair** drop-down of the **Build** panel in the **Home** tab. This tool is used to create a stair tower from one stair. You can create a stair tower by using any stair shape except spiral. To create a stair tower, you need to fulfill the condition given next.

1.  A stair must already be created in the drawing.
2.  You must have more than one level in the current project.

# CREATING GRIDS, BEAMS, COLUMNS, AND BRACES

To create grids, beams, columns, or braces, the tools are available in the **Enhanced Custom Grid** drop-down in the **Build** panel of the **Home** tab in the **Ribbon**. The tools in this drop-down are discussed next.

## Enhanced Custom Grid

This tool is available in the **Enhanced Custom Grid** drop-down. This tool is used to create column grids. When you choose this tool, the **Column Grid** dialog box will be displayed, as shown in Figure 3-32. Using the options in this dialog box, you can create two types of grids: orthogonal grids and radial grids. The methods to create these grids are discussed next.

### Creating Orthogonal Grids

To create orthogonal grids, select the **Orthogonal** radio button from the top left corner of the **Grid Layout** area; the modified **Grid Layout** area will be displayed. By default, the **Top** tab is selected in the dialog box. As a result, the grid lines will be created in the Top plane. To create a grid line, select the desired distance value from the list available on the left in the **Grid Layout** area; the preview of the grid lines distanced as per the selected value, will be displayed in the **Preview** area. Also, the parameters of the grid lines will be displayed in the table on the right in the **Grid Layout** area. Keep on clicking on the desired distance values in the list till you get the required number of grid lines. Figure 3-33 shows the **Column Grid** dialog box after selecting values from the list displayed in the left. Similarly, you can add desired number of grids on Bottom, Left, and Right plane. You can change the pattern of labeling by using the options available in the **X-Labeling** and **Y-Labeling** drop-down lists. Similarly, you can change other parameters for grids by using the options available in the dialog box. After setting all the parameters, choose the **OK** button; the grid will be displayed attached to the cursor and you will be prompted to specify the insertion point. Specify the insertion point; you will be prompted to specify the rotation angle. Specify the rotation angle; the grid will be created at the specified location. Press ENTER to exit the command or you can specify an insertion point to create another grid with the same specifications.

*Figure 3-32  The **Column Grid** dialog box*

*Figure 3-33* The **Column Grid** dialog box displayed after adding grid lines

## Creating Radial Grids

To create radial grids, select the **Radial** radio button on the top left corner in the **Grid Layout** area; the modified **Grid Layout** area will be displayed, refer to Figure 3-34. By default, the **Radial** tab is selected in the dialog box, as a result the grid lines will be created in the radial direction. To create a grid line, select the desired angle value from the list available in the left of the dialog box; preview of the grid lines, at the selected value, will be displayed in the **Preview** area. Also, the parameters of the grid lines will be displayed in the table on the right in the **Grid Layout** area. You can add the desired number of grid lines by selecting from the list. Figure 3-35 shows the **Column Grid** dialog box after selecting values from the list displayed in the left. Similarly, you can add desired number of grids in the form of an arc by selecting the **Arcs** tab. You can change the pattern of labeling by using the options available in the **X-Labeling** and **Y-Labeling** drop-down lists. Similarly, you can change other settings for grids by using the options available in the dialog box. After setting all the parameters, choose the **OK** button; the grid will be displayed attached to the cursor and you will be prompted to specify the insertion point. Specify the insertion point; you will be prompted to specify the rotation angle. Specify the rotation angle; the grid will be created at the specified location. Press ENTER to exit the command or you can specify an insertion point to create another grid with same specifications.

*Figure 3-34* *The modified* **Column Grid** *dialog box after selecting the* **Radial** *radio button*

*Figure 3-35* *The modified* **Column Grid** *dialog box after selecting values from the list*

## Custom Grid Convert

Custom Grid Convert    This tool is available in the **Enhanced Custom Grid** drop-down. This tool is used to convert the network of lines into column grids. When you choose this tool, you will be prompted to select a network of lines. The command prompt for creating a grid from lines in given next.

Select linework: *Select the network of lines that you want to convert in grid.*
Select linework: Enter
Enter label extension or [No labels] <1200>: *Specify the length of extension line from the grid end point.*
Erase selected linework? [Yes/No] <No>: *Specify whether you want to delete the network of lines or not.*

## Column Grid

Column Grid    This tool is available in the **Enhanced Custom Grid** drop-down. This tool is used to create a column grid with predefined parameters. To create a column grid, choose the **Column Grid** tool from the drop-down; you will be prompted to specify the insertion point for the grid. Specify the insertion point; you will be prompted to specify the rotation angle for the grid. Specify the rotation angle; the grid will be created at the specified insertion point. Again, you will be prompted to specify the insertion point for the grid. Press ENTER to exit the command. You can edit the parameters of a grid by using the options available in the **Properties** palette, refer to Figure 3-36. Some important options in this palette are discussed next.

*Figure 3-36* The **Properties** palette displayed on choosing the **Column Grid** tool

## Shape

This drop-down list is available in the **General** rollout of the **Properties** palette. The options in this drop-down are used to change the shape of the grid to be created. There are two options available in this drop-down: **Rectangular** and **Radial**. When the **Rectangular** option is selected, the grid to be created will be in rectangular shape. There will be horizontal and vertical grid lines in the grid. When you select the **Radial** option, the grids will be in the form of arcs and in the radial direction.

## Specify on screen

This drop-down is available in the **Dimensions** rollout of the **Properties** palette. The options in this drop-down are used to specify the method for dimensioning the column grids. There are two options available in this drop-down: **Yes** and **No**. When you choose the **Yes** option, you will be prompted to specify the corner points of the column grids in the screen. If the **No** option is chosen, then you need to specify the dimensions in the edit boxes available in the **Properties** palette.

## X - Width

This edit box is available for both rectangular shaped as well as the radial shaped grids. This edit box in the **Dimensions** rollout of the **Properties** palette gets activated only when **No** option is selected in the **Specify on screen** drop-down. This edit box is used to specify the value for width of the grid.

## Y - Depth

This edit box is available for rectangular shaped grids only. This edit box in the **Dimensions** rollout of the **Properties** palette gets activated only when **No** option is selected in the **Specify on screen** drop-down. This edit box is used to specify the value for depth of the grid.

## A - Angle

This edit box is available for radial shaped grids only. This edit box in the **Dimensions** rollout of the **Properties** palette gets activated only when the **No** option is selected in the **Specify on screen** drop-down. This edit box is used to specify the value for angle of the grid.

## Layout type

This drop-down list is available in both the **X Axis** rollout as well as the **Y Axis** rollout. There are two options available in this drop-down: **Repeat** and **Space evenly**. When the **Repeat** option is selected, the distance between the two grids is fixed and depending on the total size specified, the number of grid lines are inserted. If the **Space evenly** option is selected in the **Layout type** drop-down, the total number of grid lines will be fixed and depending on the total size of grid boundary, the distance between the two grid lines will be adjusted.

## Bay size

This edit box is available in both the **X Axis** and the **Y Axis** rollouts. This edit box is available only when the **Repeat** option is selected in the **Layout type** drop-down. You can specify the distance between two grid lines perpendicular to the X axis and Y axis by using the edit box in the respective rollout.

## Number of bays

This edit box is available in both the **X Axis** rollout as well as the **Y Axis** rollout. This edit box is available only when the **Space evenly** option is selected in the **Layout type** drop-down. You can specify the number of grid lines along X axis by using the edit box in the respective rollout.

# Column

This tool is available in the **Enhanced Custom Grid** drop-down. This tool is used to create a column with predefined parameters. To create a column, choose the **Column** tool from the drop-down; you will be prompted to specify the position of column in the screen. Select a point on the grid or specify a point in the drawing area; the column will be created at the specified point. You can change the parameters of the column by using the options available in the **Properties** palette, refer to Figure 3-37. Some of the important options in the **Properties** palette are discussed next.

## On object

This drop-down list is available in the **General** rollout. There are two options available in this drop-down list: **Yes** and **No**. When you select the **Yes** option, the column will be attached to the base object. If you move the base, the column will move accordingly. When you choose the **No** option, the column will be placed as a separate object having no link with the other objects.

*Figure 3-37  The Properties palette displayed on choosing the Column tool*

## Start offset
This edit box is available in the **Dimensions** rollout. In this edit box, you can specify the offset value for the start point of the column from the insertion point.

## End offset
This edit box is available in the **Dimensions** rollout. In this edit box, you can specify the offset value for the end point of column from the insertion point.

## Logical Length
This edit box is available in the **Dimensions** rollout. In this edit box, you can specify the value for length of the column.

## Specify roll on screen
This drop-down list is available in the **Dimensions** rollout. There are two options available in this drop-down list: **Yes** and **No**. When you select the **Yes** option, you will be prompted to specify the value for rotation of column while placing it. When you select the **No** option, you need to specify the value for rotation of column in the **Roll** edit box available in the **Properties** palette.

## Roll
This edit box is available in the **Dimensions** rollout only when the **No** option is selected in the **Specify roll on screen** drop-down list. In this edit box, you can specify the value for rotation of the column being created.

## Justify
This drop-down list is available in the **Dimensions** rollout. There are ten options available in this drop-down list to change the justification of the column.

## Justify cross-section
This drop-down list is available in the **Dimensions** rollout. There are two options available in this drop-down list: **Maximum** and **At each node**. When you select the **Maximum** option from the drop-down list, the justification method is applied only to the columns having maximum cross section. When you choose the **At each node** option from the drop-down list, the justification method is applied to every column.

## Justify components
This drop-down list is available in the **Dimensions** rollout. There are two options available in this drop-down list: **Highest priority only** and **All**. When you select the **Highest priority only** option from the drop-down list, the justification methods will be applied only to the columns having highest priority. When the **All** option is selected from the drop-down list, then the justification method is applied to all the columns.

## Custom Column

This tool is available in the **Enhanced Custom Grid** drop-down. This tool is used to create a user defined column. To create a custom column, choose the **Custom Column** tool; you will be prompted to select a closed polyline, a closed spline, a circle, or an ellipse. Select a closed entity; you will be prompted to specify the insertion point or centroid. Select a point to specify as insertion point or centroid; the **Convert to Column** dialog box is displayed, as shown in Figure 3-38. Specify the name of the column in the **New Name** edit box and choose the **OK** button from the dialog box; the custom column will be created at the specified location.

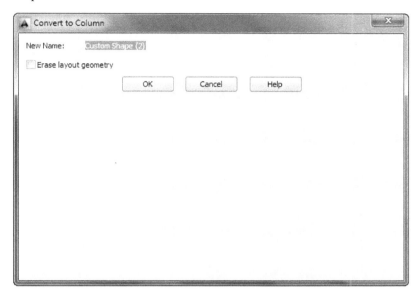

*Figure 3-38* *The Convert to Column dialog box*

You can change the properties of a custom created column by using the **Properties** palette, refer to Figure 3-39. Some important options available in the **Properties** palette after selected a custom column are discussed next.

## Bound spaces

This drop-down list is available in the **General** rollout. There are three options available in this drop-down list: **Yes**, **No**, and **By style**. When the **Yes** option is selected in the drop-down list, the created object will act as a space boundary. When the **No** option is selected in the drop-down list, the created object will be not act as a space boundary. When the **By style** option is selected in the drop-down list, the boundary conditions of the object are specified by the style of object.

*Figure 3-39  The **Properties** palette*

## Trim automatically

This drop-down list is available in the **General** rollout. There are three options available in this drop-down list: **Yes**, **No**, and **By style**. When the **Yes** option is selected in the drop-down list, the created object will get automatically trimmed to join with the connecting object. When the **No** option is selected in the drop-down list, the created object will not get automatically trimmed to join with the connecting object. When the **By style** option is selected in the drop-down list, the trimming conditions of the object are specified by the style of object.

## Member type

This drop-down list is available in the **General** rollout. There are three options available in this drop-down list: **Beam**, **Column**, and **Brace**. You can choose the **Beam** option from this drop-down list to create a custom beam. By default, the **Column** option is selected in the drop-down list. This option is used to create a column. You can create a brace by choosing the **Brace** option from the drop-down list.

The other options in the **Properties** palette have already been discussed.

# Beam

This tool is available in the **Enhanced Custom Grid** drop-down. This tool is used to create a beam. To create a beam, choose the **Beam** tool; you will be prompted to specify the start point of the beam. Specify the start point of the beam; you will be prompted to specify the end point of the beam. Specify the end point of the beam; the beam will be created with the specified parameters. You can specify the parameters of the beam in the **Properties** palette. The options available in the **Properties** palette for beam are similar to the options available for custom column. The options that are available only for a beam are discussed next.

## Array

This drop-down list is available in the **Layout** rollout. There are two options available in this drop-down list: **Yes** and **No**. The **No** option is selected by default in the drop-down list. When you select the **Yes** option, you can create multiple instances of the beam. On selecting the **Yes** option, the **Layout method** and **Number of bays** options become available in the **Layout** rollout. These options are used to specify the parameter for creating multiple instances of the beam.

## Layout method

This drop-down list is available in the **Layout** rollout. There are two options available in this drop-down list: **Space evenly** and **Repeat**. When you select the **Space evenly** option, the instances are spaced equally over the total distance. When you select the **Repeat** option, the instances are spaced according to the specified distance value specified.

## Number of bays

This edit box is available in the **Layout** rollout. In this edit box, you can specify the value for number of instances of the beam.

# Brace

This tool is available in the **Enhanced Custom Grid** drop-down. This tool is used to create a brace. To create a brace, choose the **Brace** tool; you will be prompted to specify the start point of the brace. Specify the start point of the brace; you will be prompted to specify the end point of the brace. Specify the end point of the brace; the brace will be created with the specified parameters. You can specify the parameters of the brace in the **Properties** palette. The options available in the **Properties** palette for brace are similar to the options available for custom column. The options that are available only for a brace are discussed next.

## Specify rise on screen

This drop-down list is available in the **Rise** rollout. There are two options available in this drop-down list: **Yes** and **No**. The **Yes** option is selected by default in the drop-down list. So you need to specify the value of rise on the screen while placing the braces. When you choose the **No** option, you can set the parameters related to rise by using the options available in the **Properties** palette.

## Method

This drop-down list is available in the **Rise** rollout. There are two options available in this drop-down list: **Angle** and **Distance**. On choosing the **Angle** option, you can set the value of rise by specifying angle value in the **Angle from first member** edit box.

## Distance along first member

This edit box is available in the **Rise** rollout. You can specify the value for distance of start point of the brace along the first member.

## Angle from the first member

This edit box is available only when the **Angle** option is selected in the **Method** drop-down. You can specify the value for rotation of the brace with respect to the first member.

# CREATING PRIMITIVES

You can create basic objects like box, cylinder, torus and so on by using the tools available in AutoCAD MEP. These basic objects are called primitives. The tools to create primitives are available in the **Box** drop-down of the **Build** panel. The tools available in this drop-down are discussed next.

## Box

The **Box** tool is used to create a box with specified dimensions. To create a box, choose the **Box** tool from the **Box** drop-down in the **Build** panel; you will be prompted to specify the first corner of the box. Specify the first corner of the box; you will be prompted to specify the second corner of the box. Specify the second corner of the box; you will be prompted to specify the height of the box. Specify the height of the box; you will be prompted to specify the rotation angle of the box. Specify the rotation value; the box will be created.

## Pyramid

The **Pyramid** tool is used to create a pyramid with specified dimensions. To create a pyramid, choose the **Pyramid** tool from the **Box** drop-down in the **Build** panel; you will be prompted to specify the first corner point of the pyramid. Specify the first corner point of the pyramid; you will be prompted to specify the second corner point. Specify the second corner point; you will be prompted to specify the height of the apex. Specify the height of the apex; you will be prompted to specify the rotation angle of the pyramid. Specify the value for rotation; a pyramid with specified settings will be created.

## Cylinder

The **Cylinder** tool is used to create a cylinder with specified dimensions. To create a cylinder, choose the **Cylinder** tool from the **Box** drop-down in the **Build** panel; you will be prompted to specify the insertion point of the pyramid. Specify the point in the drawing area; you will be prompted to specify the radius. Enter the radius value; you will be prompted to specify the height of the cylinder. Specify the height of the cylinder; you will be prompted to specify the rotation angle. Specify the value for rotation; a cylinder with specified settings will be created.

## Right Triangle

The **Right Triangle** tool is used to create a right triangle mass with specified dimensions. To create a right triangle, choose the **Right Triangle** tool from the **Box** drop-down in the **Build** panel; you will be prompted to specify the first corner point of the right triangle. Specify the point in the drawing area; you will be prompted to specify the second corner point of the right triangle. Specify the second corner point; you will be prompted to specify the height of the apex. Specify the height of the apex; you will be prompted to specify the rotation angle of the right triangle. Specify the value for rotation; a right triangle with specified settings will be created.

## Isosceles Triangle

The **Isosceles Triangle** tool is used to create an isosceles triangle mass with specified dimensions. The procedure to create an isosceles triangle is similar to the procedure of creating a right triangle mass.

## Cone

The **Cone** tool is used to create a cone with specified dimensions. To create a cone, choose the **Cone** tool from the **Box** drop-down in the **Build** panel; you will be prompted to specify the insertion point of the cone. Specify the point in the drawing area; you will be prompted to specify the radius. Enter the radius value; you will be prompted to specify height of the cone. Specify the height of the cone; you will be prompted to specify the rotation angle. Specify the value of rotation; a cone with specified settings will be created.

## Dome

The **Dome** tool is used to create a dome with specified dimensions. To create a dome, choose the **Dome** tool from the **Box** drop-down in the **Build** panel; you will be prompted to specify the insertion point of the dome. Specify the point in the drawing area; you will be prompted to specify the radius. Enter the radius value; you will be prompted to specify the rotation angle. Specify the value of rotation; a dome with specified settings will be created.

## Sphere

The **Sphere** tool is used to create a dome with specified dimensions. To create a sphere, choose the **Sphere** tool from the **Box** drop-down in the **Build** panel; you will be prompted to specify the insertion point of the sphere. Specify the point in the drawing area; you will be prompted to specify the radius. Enter the radius value; you will be prompted to specify the rotation angle. Specify the value of rotation; a sphere with specified settings will be created.

## Arch

The **Arch** tool is used to create an arch with specified dimensions. To create an arch, choose the **Arch** tool from the **Box** drop-down in the **Build** panel; you will be prompted to specify the first corner point of the base. Specify the point in the drawing area; you will be prompted to specify the second corner point of the base. Specify the second corner point; you will be prompted to specify the height of the arch. Specify the height of the arch; you will be prompted to specify the rotation angle of the arch. Enter the value of rotation angle; an arch will be created with the specified parameters.

## Gable

The **Gable** tool is used to create an arch with specified dimensions. To create a gable, choose the **Gable** tool from the **Box** drop-down in the **Build** panel; you will be prompted to specify the first corner point of the base. Specify the point in the drawing area; you will be prompted to specify the second corner point of the base. Specify the second corner point; you will be prompted to specify the height of the gable. Specify the height of the gable; you will be specify the rotation angle of the arch. Enter the value of the rotation angle; a gable will be created with the specified parameters.

## Barrel Vault

The **Barrel Vault** tool is used to create a barrel vault with specified dimensions. To create a barrel vault, choose the **Barrel Vault** tool from the **Box** drop-down in the **Build** panel; you will be prompted to specify the first corner point of the base. Specify the point in the drawing area; you will be prompted to specify the second corner point of the base. Specify the second corner point; you will be prompted to specify the rotation angle of the barrel vault. Enter the value of the rotation angle; a barrel vault will be created with the specified parameters.

## Drape

The **Drape** tool is used to create a drape with specified parameters. A drape is used to show contour of the site. You can create a drape with the help of polylines or polygons. To create a drape, choose the **Drape** tool from the **Box** drop-down in the **Build** panel; you will be prompted to select objects representing the contour. Select polylines or other objects representing the contour and then press ENTER; you will prompted to erase or retain the selected contours. Enter **Y** or **N** at the command prompt; you will be prompted to generate a regular or irregular mesh. Enter **Y** or **N** at the command prompt. On specifying **N** at the command prompt, you will be prompted whether to generate a rectangular mesh. On entering **Y** at the command prompt; you will be prompted to specify the first corner of the rectangle to create a mesh. Specify a point in the drawing area; you will be prompted to specify the opposite corner of the rectangular mesh. Specify the opposite corner of the mesh; you will be prompted to specify the number of subdivisions along X direction. Enter the desired number of subdivisions at the command prompt; you will be prompted to specify the number of subdivisions along the Y direction. Enter the number of subdivisions at the command prompt; you will be prompted to specify the thickness value of the base. Enter the value of thickness at the command prompt; the drape with the selected contours will be created.

# TUTORIALS

## Tutorial 1

In this tutorial, you will create a model of a building, as shown in Figure 3-40. The plan of the building is given in Figure 3-41.                                **(Expected time: 30 min)**

**Figure 3-40** *Model of a building*

**Figure 3-41** *Plan view of the building*

Examine the model to determine the number of features in it. The model consists of three features, refer to Figure 3-40.

The following steps are required to complete this tutorial:

a.   Create a project file with the name Office.
b.   Create a new construct with the name C03_tut01.
c.   Create walls by using the reference of line diagram, refer to Figures 3-41 and 3-42.
d.   Create doors according to the plan, refer to Figures 3-41 and 3-43.

## Starting a New Project File

1.   Start AutoCAD MEP by using AutoCAD MEP 2014 - English (Global) icon from the desktop and choose **New > Project** from the **Application** menu; the **Project Browser** is displayed.

2.   Choose the **New Project** button available at the bottom of the **Project Browser**; the **Add Project** dialog box is displayed.

3.   Enter the project name as **Office** in the **Project name** field. Also, enter the project number as **0001** and specify the description about the project as desired in their respective fields. Make sure that the **Create from template project** check box is selected.

4.   Choose the Browse button and select the file *Commercial Template Project (Metric).apj* from the *C:\ProgramData\Autodesk\MEP 2014\enu\Template\Commercial Template Project (Metric)* directory and then choose the **Open** button; the path of the selected file will be added to the edit box displayed below the **Create from template project** check box.

5.   Choose the **OK** button; the newly created project file name is displayed in the left pane of the **Project Browser**.

6.   Choose the **Close** button from the **Project Browser**; the **Project Navigator** palette is displayed in the drawing area and the new project is activated. By default, the **Constructs** tab is active in the **Project Navigator**.

## Adding a New Construct

1.   Choose the **Add Construct** button available at the bottom of the **Project Navigator**; the **Add Construct** dialog box is displayed.

2.   Click in the **Name** field, specify the name of construct as *C03_tut01* and choose the **OK** button; the *C03_tut01* drawing file is opened.

## Creating Walls

Now, you will create walls. Before that choose the **Architecture** option from the **Workspace Switching** flyout.

1.   Choose the **Wall** tool from the **Wall** drop-down available in the **Build** panel of the **Home** tab; you are prompted to specify the start point of the wall. Also, the corresponding **Properties** palette is displayed.

2. Specify width, base height, and justify as **254**, **4000**, and **Center** respectively in their respective fields in the **Dimensions** rollout of the **Properties** palette.

3. Click to specify the start point of the wall and then create the walls, as shown in Figure 3-42. For dimensions, refer to Figure 3-41.

## Creating Doors

Now, you will create doors. For dimensions, refer to Figure 3-41.

1. Choose the **Door** tool from the **Door** drop-down available in the **Build** panel of the **Home** tab; you are prompted to select a wall or a grid assembly. Also, the **Properties** palette is displayed.

*Figure 3-42    Walls created*

2. Specify width and height as **1500** and **3500** respectively, in their respective fields in the **Dimensions** rollout of the **Properties** palette. Also, select the **Center** option from the **Position along wall** field in the **Location** rollout of the **Properties** palette.

3. Click at the required locations to place the doors D1, refer to Figure 3-41.

4. Now, specify width as **2000** in the **Width** field available in the **Dimensions** rollout of the **Properties** palette and create the main gate. For positioning the gates, refer to Figure 3-41.

After creating the doors, the drawing is displayed, as shown in Figure 3-43.

## Creating Windows

Now, you will create windows with the specifications given in Figure 3-41.

1.  Choose the **Window** tool from the **Window** drop-down available in the **Build** panel of the **Home** tab; you are prompted to select a wall or a grid assembly. Also, the **Properties** palette with corresponding options is displayed.

2.  Specify the width and height of window as **2000** and **1510** respectively, in their corresponding fields in the **Dimensions** rollout of the **Properties** palette.

3.  Click at the required locations to create the windows, refer to Figure 3-41.

*Figure 3-43*   *The drawing after creating gates*

The final model in the **SW Isometric** view and the **Conceptual** display style is shown in Figure 3-44.

*Figure 3-44*   *Model of a building*

### Saving the Drawing file

1. Choose **Save** from the **Application** menu to save the drawing file with the name *c03_tut1*.

## Tutorial 2

In this tutorial, you will create a model of a building, as shown in Figure 3-45. The plan of the building is given in Figure 3-46. **(Expected time: 30 min)**

***Figure 3-45*** *Model of a building*

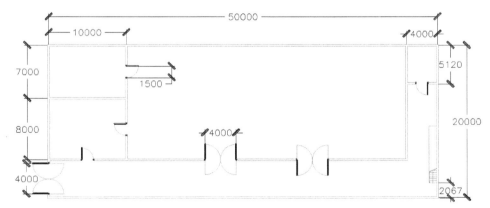

***Figure 3-46*** *The plan of the building*

The following steps are required to complete this tutorial:

a. Create a new project file.
b. Create a new construct with the name C03_tut02.
c. Create walls by using the reference of line diagram, refer to Figure 3-47.
d. Create doors according the plan, refer to Figure 3-48.

e.  Create the roof according to the plan, refer to Figure 3-49.

f.  Create the stairs, refer to Figure 3-50.

After starting the AutoCAD MEP session, the first task is to create a project.

## Starting a New Project File

1.  Start AutoCAD MEP by using AutoCAD MEP 2014 - English (Global) icon from the desktop, if not started and then choose **New** > **Project** from the **Application** menu; the **Project Browser** is displayed.

2.  Choose the **New Project** button available at the bottom of the **Project Browser**; the **Add Project** dialog box is displayed.

3.  Enter the project name as **Machining Plant** in the **Project name** field. Also, enter the project number and description in their respective fields. Select the *Commercial Template Project (Metric).apj* as template for the project.

4.  Choose the **OK** button; the new created project file name is displayed in the left pane of the **Project Browser**.

5.  Choose the **Close** button from the **Project Browser**; the **Project Navigator** palette is displayed in the drawing area and the new project is activated.

## Adding a New Construct

1.  Select the **Constructs** tab from the **Project Navigator**. Choose the **Add Construct** button available at the bottom of the **Project Navigator**; the **Add Construct** dialog box is displayed.

2.  Click in the **Name** field, specify the name of construct as *C03_tut02* and choose the **OK** button; the *C03_tut02* drawing file is opened.

## Creating Walls

Now, you need to create walls. Make sure that the **Architecture** workspace is active.

1.  Choose the **Wall** tool from the **Wall** drop-down available in the **Build** panel of the **Home** tab; you are prompted to specify the start point of the wall. Also, the **Properties** palette is displayed.

2.  Specify width, base height, and justify as **254**, **4500**, and **Center** respectively in their respective fields in the **Dimensions** rollout of the **Properties** palette.

3.  Click to specify the start point of the wall and then create the walls, as shown in Figure 3-47. For dimensions, refer to Figure 3-46.

## Creating Doors

Now, you need to create doors. For dimensions, refer to Figure 3-46. Make sure that the **Architecture** workspace is active.

1. Choose the **Door** tool from the **Door** drop-down available in the **Build** panel of the **Home** tab; you are prompted to select a wall or a grid assembly. Also, the **Properties** palette with the corresponding options is displayed.

*Figure 3-47* *Walls created using reference of line diagram*

2. Specify width and height as **1500** and **3500** respectively, in their respective fields in the **Dimensions** rollout of the **Properties** palette. Also, select the **Center** option from the **Position along wall** field in the **Location** rollout of the **Properties** palette.

3. Click at the required locations to create doors, refer to Figure 3-46.

   Now, you need to create double-dhung doors. These doors are not available by default, so you need to create a new door style first.

4. Choose the **Style Manager** tool from the **Style & Display** panel in the **Manage** tab of the **Ribbon**; the **Style Manager** is displayed.

5. Click on the **+** sign adjacent to **Architectural Objects**; the list of architectural objects is displayed.

6. Right click on **Door Styles** from the list and select the **New** option from the shortcut menu displayed; a new style will be added to the **Door Styles** category and various options related to the new style are displayed in the right area.

7. Choose the **Design Rules** tab from the right area if not chosen by default. Select the **Double-Dhung** option from the **Door Type** area and choose the **OK** button.

   After creating new style, you will use this style for creating the main gate and the entry to the machining zone.

8. Select the **Door** tool from the **Door** drop-down available in the **Build** panel of the **Home** tab; you are prompted to select a wall or a grid assembly. Also, the **Properties** palette with the corresponding options is displayed.

9.  Specify width and height as **4000** and **3500** respectively, in their respective fields in the **Dimensions** rollout of the **Properties** palette. Select the **Unconstrained** option from the **Position along wall** field in the **Location** rollout of the **Properties** palette.

10. Choose the **New Style** option from the **Style** drop-down list in the **General** rollout and position the doors at the locations shown in the Figure 3-46.

    After creating doors, the drawing is displayed as shown in Figure 3-48.

*Figure 3-48*  *The drawing after creating all doors*

## Creating Slab

Now, you need to create the slab with the specifications shown in Figure 3-46.

1.  Choose the **Slab** tool from the **Roof Slab** drop-down available in the **Build** panel of the **Home** tab; you are prompted to specify the start point of the slab.

2.  Select one of the corner point of the building displayed in Figure 3-49; you are prompted to specify the next point for slab.

3.  Select rest of the corner points as shown in Figure 3-49.

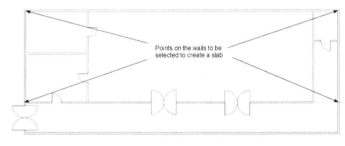

*Figure 3-49*  *Corner points for slab*

## Creating Stairs

Now, you will create stairs attached to the slab.

1.  Choose the **Stair** tool from the **Stair** drop-down available in the **Build** panel of the **Home** tab; you are prompted to specify the flight start point. Also, the **Properties** palette is displayed with the options relevant to stairs.

2.  Specify width, height, justification, and termination as **1000**, **4500**, **Right**, and **Landing** respectively, in their respective fields in the **Properties** palette. Also, choose **Straight** from the **Shape** drop-down list in the **General** rollout.

3.  Select a starting point for stairs and align it with the adjacent wall, refer to Figure 3-50.

*Figure 3-50*  *Starting point of the stairs*

Now, you will create a cut on the slab to allow passage through the stairs to the top of the roof.

## Creating Cut in the Slab

1.  Choose the **Rectangle** tool from the **Rectangle** drop-down available in the **Draw** panel of the **Home** tab; you are prompted to specify the first corner point. Specify the corner points of rectangle, as shown in Figure 3-51.

*Figure 3-51*  *Rectangle to be created for cut in the slab*

2.  Select the slab and choose the **Trim** tool from the **Modify** panel in the **Slab** contextual tab; you are prompted to select a trimming object like a polygon or a solid object.

3. Select the rectangle created for the cut and press ENTER; you are prompted to specify the side to be deleted.

4. Click inside the rectangle, the area covered by the rectangle is trimmed. Press ESC to exit the tool. Now, hide the rectangle geometry created for trimming,

   Isometric view of the drawing after creating the cut is shown in Figure 3-52.

*Figure 3-52*   *Drawing after creating the cut in the slab*

## Creating Railing on the Stairs

1. Choose the **Railing** tool from the **Stair** drop-down available in the **Build** panel of the **Home** tab; you are prompted to specify the start point of the railing.

2. Choose **Attach** and then **Stair** option from the command prompt; you are prompted to select a stair.

3. Select the stair created earlier; a railing with standard settings is created in the drawing area.

4. Deselect all the entities, select the railing, and invoke the **Properties** palette.

5. Enter **900** in the **Side offset** field in the **Location** rollout of the **Properties** palette.

   After creating all the components, the final drawing (in conceptual view) is displayed, as shown in Figure 3-53.

## Saving the Drawing File

1. Choose **Save** from the **Application** menu to save the drawing file.

**Figure 3-53**   *The final drawing*

## Self-Evaluation Test

Answer the following questions and then compare them to those given at the end of this chapter:

1.   You cannot create a curved wall using the **Wall** tool. (T/F)

2.   A curtain wall is a non-structural wall. (T/F)

3.   Using the **Stair** tool, you can create only straight stairs. (T/F)

4.   A _____ is an opening in a wall or door to facilitate the passage of air and light.

5.   The _____ tool is used to create a segment of roof which is not connected with other entities.

6.   A _____ window can be created on a cornered wall.

## Review Questions

Answer the following questions:

1.   Using the options in the **Style Manager**, you can create custom doors. (T/F)

2.   Spiral stairs can be created only in the clockwise direction. (T/F)

3.   The **Railing** tool can be used to attach railing to a stair. (T/F)

4.   On selecting a slab, the _____ contextual tab gets activated in the **Ribbon**.

5.  The _____ tool is used to convert the network of lines into column grids.

# Exercise

## Exercise 1

Create the model of building, as shown in Figure 3-54. The plan view of the building is shown in Figure 3-55.                                    **(Expected time: 45 min)**

*Figure 3-54* *Model for Exercise 1*

*Figure 3-55* *Plan of the building*

## Answers to Self-Evaluation Test
**1.** F, **2.** T, **3.** F, **4.** window, **5. Roof Slab**, **6.** cornered

# Chapter 4

# Creating the HVAC System

## Learning Objectives

*After completing this chapter, you will be able to:*

- *Use HVAC equipment*
- *Change basic settings of an HVAC system*
- *Configure ductwork options*
- *Route the duct line*
- *Create ducts*

# INTRODUCTION

In this chapter, you will learn about the use of mechanical equipment that are required for creating an HVAC system. An HVAC system is used to maintain desired environmental conditions in a specific area. It consists of the heating, ventilation, and air conditioning systems. To create an HVAC system, you need to know the heating load of the desired area by using thermodynamics, fluid mechanics, and heat transfer principles. These load settings are used to determine the capacity of equipment to be added for HVAC. To start working in the **HVAC** workspace, choose the **HVAC** option from the **Workspace Switching** flyout in the **Drawing Status Bar**. The equipment available in the **HVAC** workspace of AutoCAD MEP are discussed next.

# EQUIPMENT

For creating an HVAC system, you need to add equipment related to the system to the building structure. All the equipment that can be added while working in the **HVAC** workspace are available in the **Equipment** drop-down of the **Build** panel in the **Home** tab of the **Ribbon**, refer to Figure 4-1. The tools available in this drop-down are discussed next.

## Air Handler

Air Handler is an equipment or a device which is used to circulate air in a specific area. The quality of air is also controlled by Air Handler. There are two types of Air Handlers available in AutoCAD MEP: Modular Air Handler and Packaged Air Handler. A Modular Air Handler consists of various components that together make an Air Handler. Some of the important components in a Modular Air Handler are AHU Coils, AHU Economizer, AHU Fans, AHU Filters, AHU Inspection Modules, and AHU Mixing Boxes. A Packaged Air Handler is a closed unit consisting of all the components available in the Modular Air Handler but in an interconnected manner. To add an Air Handler in the drawing, choose the **Air Handler** tool from the **Equipment** drop-down available in the **Build** panel of the **Home** tab in the **Ribbon**; the **Add Multi-view Parts** dialog box will be displayed, as shown in Figure 4-2. Select the AHU Large Coils component

*Figure 4-1 The **Equipment** drop-down*

from the AHU Coils sub-category of the Modular Air Handling Unit Components category of Air Handling Units equipment. Also, the preview of the selected component is displayed in the right of the dialog box. The blue arrows in the Preview area indicate the flow of fluid through the selected component. Various components in the Modular Air Handling Unit Components category are discussed next.

## AHU Coils

The AHU Coil is a component of Modular Air Handler which is used to cool or heat the air coming through the duct. There are four ports of AHU Coil component: Duct inlet port, Duct outlet port, Heat Exchanging fluid inlet port, and Heat Exchanging fluid outlet port. To add an AHU Coil in the drawing, click on the **+** sign adjacent to AHU Coil in the Modular Air Handling Unit Components category of Air Handling Units equipment; three types of AHU Coil will be displayed below the AHU Coil sub-category.

*Figure 4-2  The **Add Multi-view Parts** dialog box*

You can choose any of the three options: AHU Large Coil, AHU Medium Coil, and AHU Small Coil, as per the requirement. After choosing the desired option, select the required size from the **Part Size Name** drop-down list available below the Preview area. After selecting the respective size for AHU coil, click in the drawing area; the AHU coil will be placed at the selected position and a compass will be displayed below the AHU coil. Orient the AHU coil at the desired angle by using the compass. Figure 4-3 shows an annotated AHU Coil.

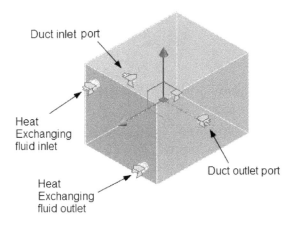

*Figure 4-3  An AHU Coil*

## AHU Economizer

The AHU Economizer is a component of the AHU system which is used to increase the efficiency of an AHU system by using natural air. In an Economizer, natural air is used to heat or cool the air in a specific area. For example, if the temperature inside a room is higher than the temperature outside, then the economizer draws air from outside and mixes with the inside air to create an ambient temperature. To add an AHU Economizer in the drawing, click on the **+** sign adjacent to AHU Economizer in the Modular Air Handling Unit Components category of Air Handling Units equipment. On doing so, the standard AHU economizer will be displayed below the AHU Economizer sub-category. After choosing the required option from the available economizers, select the required size from the **Part Size Name** drop-down list which is available below the Preview area. After selecting the size, click in the drawing area; the AHU Economizer will be placed at the selected position and a compass will be displayed below the AHU Economizer. Orient the AHU Economizer at the desired angle by using the compass. Figure 4-4 shows an annotated AHU Economizer.

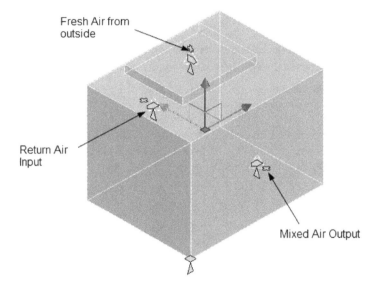

*Figure 4-4  An AHU Economizer*

## AHU Fan

The AHU Fan is a component of the AHU system which forces the air to flow through the duct. To add an AHU Fan in the drawing, click on the **+** sign adjacent to AHU Fans in the Modular Air Handling Unit Components category of Air Handling Units equipment. On doing so, three types of AHU Fans will be displayed below the AHU Fans sub-category. You can choose any of the three options: AHU Fan Modules Front Discharge Up, AHU Fan Modules Rear Discharge Up, and AHU Fan Modules Side Discharge as per the requirement. After choosing the desired option, select the required size from the **Part Size Name** drop-down list available below the Preview area. After selecting the required size for AHU Fan, click in the drawing area; the AHU Fan will be placed at the selected position and a compass will be displayed below the AHU Fan. Orient the AHU Fan at the desired angle by using the compass. Figure 4-5 shows an annotated AHU Fan Modules Side Discharge.

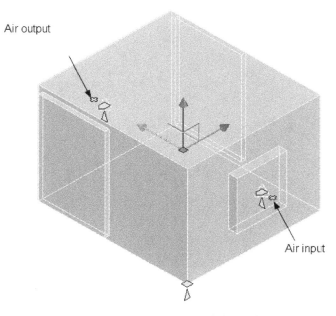

*Figure 4-5* *An AHU Fan Modules Side Discharge*

## AHU Filter
The AHU Filter is a component of the AHU system which filters the air coming from the input line. After filtering the air, it is passed to the AHU Inspection Module. To add an AHU Filter in the drawing, click on the **+** sign adjacent to AHU Filters in the Modular Air Handling Unit Components category of Air Handling Units equipment; various types of AHU Filters will be displayed below the AHU Filters sub-category. You can choose any of them as required. After choosing the required option, select the required size from the **Part Size Name** drop-down list available below the Preview area. After selecting the required size for the AHU Filter, click in the drawing area; the AHU Filter will be placed at the selected position and a compass will be displayed below the AHU Filter. Orient the AHU Filter at the desired angle by using the compass. Figure 4-6 shows an annotated AHU Bag Filter.

## AHU Inspection Modules
The AHU Inspection Module is a component of the AHU system which is used to check the quality of air being spread in the desired area. To add an AHU Inspection Module in the drawing, click on the **+** sign adjacent to AHU Inspection Module in the Modular Air Handling Unit Components category of Air Handling Units equipment; three types of AHU Inspection Modules will be displayed below the AHU Inspection Modules sub-category. You can choose any of them as required. After choosing the required option, select the required size from the **Part Size Name** drop-down list available it. After selecting the required size, click in the drawing area; the AHU Inspection Module will be placed at the selected position and a compass will be displayed below the AHU Inspection Module. Orient the AHU Inspection Module at the desired angle by using the compass. An AHU Inspection Module has two ports, one for inlet of air and other for outlet of air.

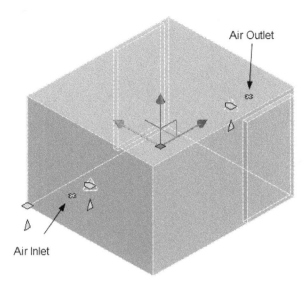

*Figure 4-6* *An AHU Bag Filter*

## AHU Mixing Boxes

The AHU Mixing Box is a component of the AHU system that mixes the returning air with the outside air as per the requirement. To add an AHU Mixing Box in the drawing, click on the **+** sign adjacent to AHU Mixing Box in the Modular Air Handling Unit Components category of Air Handling Units equipment. On doing so, standard available types of AHU Mixing Boxes will be displayed below the AHU Inspection Modules sub-category. You can choose any of the given types as required. After choosing the required option, select the required size from the **Part Size Name** drop-down list available below the Preview area. After selecting the required size, click in the drawing area; the AHU Mixing Box will be placed at the selected position and a compass will be displayed below the AHU Mixing Box. Orient the AHU Mixing Box at the desired angle by using the compass. An AHU Mixing Box has three ports, one for inlet of returning air, one for inlet of fresh air, and one for outlet of mixed air.

## Air Terminal

Air Terminal is a vent through which the fresh air is diffused in a specific area. To add a vent in the drawing, choose the **Air Terminal** tool from the **Equipment** drop-down in the **Build** panel of the **Home** tab; the **Add Multi-view Parts** dialog box will be displayed with the options for Air Terminals, as shown in Figure 4-7. There are three categories of an Air Terminal: Diffuser, Grill, and Register. Click on the **+** sign adjacent to the desired category in the dialog box; various options related to that category will be displayed below that category. Select the required option from the displayed options; different sizes for the selected option will be displayed in the **Part Size Name** drop-down list. Select the required size from the list and then click in the drawing area to place the air terminal. On placing the air terminal, you will be prompted to specify the rotation angle by using the compass displayed below it. Specify the rotation angle either by entering the value at the command prompt or dynamically specify the value by clicking on the screen. On doing so, the air terminal will be placed at the desired point and angle. Close the dialog box by choosing the **Close** button after adding the required number of air terminals.

*Figure 4-7* *The* ***Add Multi-view Parts*** *dialog box with options for Air Terminals*

**Note**
*You can specify the flow rate for the selected air terminal by using the* ***Flow (Each Terminal)*** *edit box available in the* ***Flow*** *tab of the* ***Add Multi-view Parts*** *dialog box. The unit for specifying the value should be liter/sec.*

# Fan

Fan is an equipment that creates a current of air. To add a fan in the drawing, choose the **Fan** tool from the **Equipment** drop-down in the **Build** panel of the **Home** tab; the **Add Multi-view Parts** dialog box will be displayed with the options related to fans, as shown in Figure 4-8. Select the required fan from the list; various sizes for the selected fan will be displayed in the **Part Size Name** drop-down list available below the Preview area. Select the required size from the list and click in the drawing area; the selected fan will be placed at the specified location and a compass will be displayed below the fan in the drawing area. Using this compass, you can specify the value of rotation dynamically.

*Figure 4-8*  *The* **Add Multi-view Parts** *dialog box with options for fans*

## Damper

Damper is an HVAC equipment that is used to regulate the flow of air in an HVAC system. To add a damper in the system, choose the **Damper** tool from the **Equipment** drop-down in the **Build** panel of the **Home** tab; the **Add Multi-view Parts** dialog box will be displayed with the options for dampers, as shown in Figure 4-9. There are four categories of dampers: Balancing Dampers, Fire Dampers, Rectangular Curtain Fire Shield Damper, and Rectangular Smoke Shield Damper PTC. The Balancing Damper is used to control the amount of air flowing through the duct. The Fire Damper is used to restrict the flow of fire through the duct, in case the building catches a fire. The Rectangular Curtain Fire Shield Damper is also used as a shield against fire but this type of fire dampers are in the form of curtains. The Rectangular Smoke Shield Damper PTC is used as a shield against smoke. PTC is a type of smoke shield damper which stands for Proportional Torque Control. Click on the **+** sign adjacent to the required category; a list with relative dampers will be displayed. Select the required damper from the list; various sizes for the selected damper will be displayed in the **Part Size Name** drop-down list available below the Preview area. Select the required size from the list and click in the drawing area; the selected damper will be placed at the specified location and a compass will be displayed below the damper in the drawing area. Using this compass, you can specify the value of rotation dynamically.

*Figure 4-9* *The **Add Multi-view Parts** dialog box with options for dampers*

# VAV Unit

A VAV unit is an equipment used to vary the speed of air flowing through the duct. The temperature of air to be supplied to VAV unit is constant, therefore the air flow is regulated to meet the user requirements. To add a VAV Unit in the HVAC system, choose the **VAV Unit** tool from the **Equipment** drop-down in the **Build** panel of the **Home** tab; the **Add Multi-view Parts** dialog box will be displayed with the options for VAV unit, as shown in Figure 4-10. Select the required VAV unit from the VAV Units category; the standard sizes available for the selected VAV unit will be displayed in the **Part Size Name** drop-down list. Choose the required size from the drop-down list and click in the drawing area to place the VAV unit; a compass will be displayed attached to the VAV unit for specifying the angle. Specify the rotation angle and then choose the **Close** button from the **Add Multi-view Parts** dialog box.

# Equipment

Using the **Equipment** tool, you can add any equipment available in AutoCAD MEP to the structure. When you choose the **Equipment** tool from the **Equipment** drop-down of the **Build** panel of the **Home** tab, the **Add Multi-view Parts** dialog box will be displayed, as shown in Figure 4-11. Click on the **+** sign adjacent to the Mechanical category in the dialog box; the mechanical equipment available in AutoCAD MEP will be displayed in a tree structure, refer to Figure 4-12.

*Figure 4-10* The **Add Multi-view Parts** *dialog box with options for VAV unit*

*Figure 4-11* The **Add Multi-view Parts** *dialog box*

*Figure 4-12  The **Add Multi-view Parts** dialog box with mechanical equipments*

Click on the **+** sign adjacent to the required component; various options related to the selected component will be displayed in a list. Select the required option from the list and then click in the drawing area to place the component. A compass will also be displayed below the component. Using this compass, you can rotate the component at the required angle. After specifying the angle, choose the **Close** button from the **Add Multi-view Parts** dialog box to exit the dialog box.

# DUCT LINE

Duct line is the circuit of ducts inserted in a particular area to create desired air and temperature conditions. Before creating a duct line, you must know about the specifications such as size of duct, air flow rate in the duct, number of branches required, and the size of duct in the branches. After entering these specifications, you will create the path of duct line. The methods of sizing the duct and creating a duct line are discussed next.

## Sizing the Duct Line

In AutoCAD MEP, the duct size of main duct line is dependent on two parameters: the total amount of air to be circulated which is controlled by **Flow Rate** variable and the velocity of air in the duct which is controlled by **Velocity** variable. These two variables are discussed next.

### Flow Rate

You can change the value of this variable by using the **Flow Rate** edit box available in the **Sizing** rollout of the **Properties** palette which is displayed on choosing the **Duct** tool. Flow rate is the total volume of air that is to be provided in a specific area. Enter the flow rate value in the **Flow Rate** edit box in the **Properties** palette; the value of friction and velocity will be calculated according to the rules specified in the design properties of duct type.

**Velocity**

Velocity of air flowing through a duct is a dependent variable. You cannot control the value of this variable directly. This variable is dependent on the duct size. If you increase the duct size, the value of velocity will automatically decrease and also the value of friction will decrease. If you decrease the size of duct, the value of velocity and friction will increase. In other words the duct size, you can control the value of velocity. Note that the value of velocity and friction also changes according to the option selected in the **System** drop-down list in the **BASIC** rollout.

## Routing the Duct Line

Routing is the process of creating a duct line in a layout. Routing of duct involves many factors such as position of handling units, position of diffusers, and layout of building. There are some major points that are to be considered while routing the duct line that are discussed next.

1.  Duct line should be routed in such a way that it does not pass through an area that is having a large temperature difference from the air inside the duct.
2.  Rise/Slope of duct must be determined after considering the temperature of air inside the duct.
3.  Duct routes must be as straight as possible.
4.  Duct must be larger at the opening point to the air filter entry and at the returning point of the unit.
5.  Always try to route through either the basement or the attic so that you can support the ducts by using the duct hangers.
6.  It is always good to position the air terminals and air handling unit first before routing the duct line.
7.  For a vertical run, the duct must be fastened to a wall.

# DUCT

Duct is an object which is used to deliver or remove air from an HVAC system. There are two types of ducts available in AutoCAD MEP: Duct and Flex Duct. Tools to create these ducts are available in the **Duct** drop-down in the **Build** panel of the **Home** tab in the **Ribbon**. These tools are discussed next.

## Duct

This tool is used to create a rigid duct in the HVAC system. This duct can be rectangular, oval, round, or custom shaped. To create a rigid duct, choose the **Duct** tool from the **Duct** drop-down in the **Build** panel of the **Home** tab in the **Ribbon**; the **Properties** palette will be displayed, as shown in Figure 4-13. The options available in the **Properties** palette are discussed next.

### Description

This field is available in the **General > BASIC** rollout in the **Properties** palette of the **Design** tab. When you click in this field, the **Description** dialog box will be displayed, as shown in Figure 4-14. You can enter description about the selected object in the text box available below the **Edit the description for this object** option. Choose **OK** to exit the dialog box.

*Figure 4-13   The **Properties** palette displayed on selecting the **Duct** tool*

## System

This drop-down list is available in the **General > BASIC** rollout in the **Properties** palette with the **Design** tab selected. This drop-down list contains various options for system definitions of the duct.

*Figure 4-14   The **Description** dialog box*

## Routing Preferences

This drop-down list is available in the **BASIC > Dimensions** rollout. The options available in this drop-down list are used for specifying the routing style of duct. By default, the **Standard** option is selected in this drop-down list.

## Shape

This drop-down list is available in the **BASIC > Dimensions** rollout. This drop-down list contains options to change the shape of duct. There are four options available in this drop-down list: **Oval**, **Rectangular**, **Round**, and **Undefined**. Generally, round and rectangular ducts are used in industries.

## Width

This edit box is available in the **Dimensions** rollout when the **Rectangular** or **Oval** option is selected in the **Shape** drop-down list. You can specify the value of width of the duct by using this edit box. When you click on the inverted triangle adjacent to this edit box, a drop-down list is displayed. You can select the standard sizes available for width from this edit box. The **Calculate Size** toggle button is available in the left of this edit box. When you choose this button, the width of duct is automatically calculated on the basis of flow rate, friction, velocity, and so on.

## Height

This edit box is available in the **Dimensions** rollout when the **Rectangular** or **Oval** option is selected from the **Shape** drop-down list. Using this edit box, you can specify the height of duct.

## Diameter

This edit box is available in the **Dimensions** rollout when the **Round** option is selected from the **Shape** drop-down list. Using this edit box, you can specify the diameter of duct. The **Calculate Size** toggle button is available in the left of this edit box. When you choose this button, the diameter of duct is automatically calculated on the basis of parameters such as flow rate, friction, and velocity.

## Flow Rate

Using this edit box, you can specify the rate of flow of fluid through the duct. The flow rate is specified in liters per second.

## Friction

This field is available in the **Dimensions > Sizing** rollout. This field shows the value of friction acting against the flow of fluid through the duct. The value in this field is controlled by **Duct System Definitions** section in the **Style Manager**.

## Velocity

This field is available in the **Dimensions > Sizing** rollout. This field shows the value of velocity of fluid passing through the duct. The value in this field is also controlled by **Duct System Definitions** section in the **Style Manager**.

## Specify Cut Length

This drop-down list is available in the **Dimensions > Sizing** rollout. There are two options available in this drop-down list: **Yes** and **No**. By default, the **No** option is selected in this drop-down list. On selecting the **Yes** option, you can specify the cut length in the respective field.

## Cut Length

This edit box is available in the **Dimensions > Sizing** rollout only when the **Yes** option is selected from the **Specify Cut Length** drop-down list. You can specify the value of cut length in this edit box. Cut length is the maximum length of a piece of duct that is available for creating duct line.

## Elevation

This edit box is available in the **BASIC > Placement** rollout. Using this edit box, you can specify the value of elevation of duct from the ground.

## Start Elevation

This field is available in the **BASIC > Placement** rollout. The value specified in this field is the same as the one specified in the **Elevation** edit box.

## Justify

This drop-down list is available in the **Placement > Justification** rollout. Various options are available in this drop-down list to change the justification of duct.

## Horizontal Offset

This edit box is available in the **Placement > Justification** rollout. Using this edit box, you can specify the horizontal distance value by which the duct is placed from the selected justification point.

## Vertical Offset

This edit box is available in the **Placement > Justification** rollout. Using this edit box, you can specify the vertical distance value by which the duct is placed from the selected justification point.

## Bend Method

This drop-down list is used to define the methods of bending applied on a bend in the duct system. There are three options available in this drop-down list: **Elbow**, **Offset**, and **Transition - Offset**. The **Elbow** option is selected by default.

## Bend Angle

This drop-down list is used to specify the angle of bend that is to be applied on any bend while creating a duct.

## Connection details

This button is available in the **ADVANCED** rollout. When you choose this button, the **Connection Details** dialog box will be displayed, as shown in Figure 4-15. There are two rollouts available in the dialog box that contain the details of both inlet and outlet connections.

## Lining thickness

This edit box is available in the **ADVANCED** rollout. Using this edit box, you can specify the value of thickness of liner in the duct.

*Figure 4-15* The **Connection Details** *dialog box*

## Insulation thickness

This edit box is also available in the **ADVANCED** rollout and is used to specify the thickness of insulation applied on the duct.

## Slope format

The options in this drop-down list are used to specify the format in which the slope value will be entered. There are four options available in this drop-down list: **Angle (Decimal Degrees)**; **Percentage, 100%=45 degree**; **Percentage, 100%=90 degree**; and **Rise/Run (Meters/Meters)**.

## Slope

This edit box is available in the **ADVANCED > Routing Options** rollout of the **Properties** palette. You can specify the value of slope in this edit box. The format in which the value will be specified in this edit box depends on the option selected in the **Slope format** drop-down list. If you change the slope value while adding more ducts, the fitting at joints will adjust automatically to provide the changed slope.

## Branch fitting

This drop-down list is available in the **ADVANCED > Routing Options** rollout of the **Properties** palette. There are two options in this drop-down list: **Tee** and **Takeoff**. When the **Tee** option is selected in the drop-down list, then the branches to be created will be joined to the main duct through a Tee fitting. When the **Takeoff** option is selected in the drop-down list, the branches will be joined to the main duct through the Takeoff fitting.

## Terminal-duct connection

This drop-down list is available in the **ADVANCED > Routing Options** rollout of the **Properties** palette. There are three options available in this drop-down list: **Flexible**, **Elbow with Rigid Duct**, and **Extended Duct**.

## Fitting Settings

This option is available in the **ADVANCED > Routing Options** rollout of the **Properties** palette. When you click on this option, the **Fitting Settings** dialog box will be displayed, as shown in Figure 4-16. The options in this dialog box are used to change the type of fitting to be used while creating the branches in the duct.

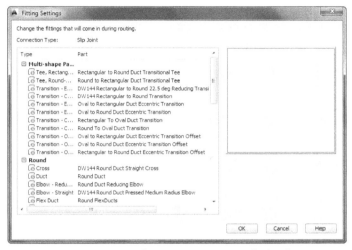

*Figure 4-16*  *The* ***Fitting Settings*** *dialog box*

## Preferences

This option is available in the **ADVANCED > Routing Options** rollout of the **Properties** palette. When you click on this option, the **Duct Layout Preferences** dialog box will be displayed, as shown in Figure 4-17. The options in this dialog box are used to define the layout preferences for creating a duct. Using the options in this dialog box, you can set the preferences for variables such as Slope, Elevation, Label style, Flow arrow style, and Elbow layout.

## Style

This drop-down list is available in the **ADVANCED > Labels and Flow Arrows > Labels** rollout of the **Properties** palette. The options in this drop-down list are used to apply the label styles to duct.

## Layout method

This drop-down list is available in the **ADVANCED > Labels and Flow Arrows > Labels** rollout of the **Properties** palette only when **NONE** is not selected from the **Style** drop-down list. There are two options available in this drop-down list: **By distance** and **By quantity**. These options are used to specify the layout method of ducts.

*Figure 4-17  The **Duct Layout Preferences** dialog box*

## Number of Labels

This edit box is available in the **ADVANCED > Labels and Flow Arrows > Labels** rollout of the **Properties** palette only when **NONE** is not selected from the **Style** drop-down list and **By quantity** is selected from the **Layout method** drop-down list. Using this edit box, you can specify total number of labels applied on a segment duct.

## Distance between

This edit box is available in the **ADVANCED > Labels and Flow Arrows > Labels** rollout of the **Properties** palette only when **NONE** is not selected from the **Style** drop-down list and **By distance** is selected from the **Layout method** drop-down list. Using this edit box, you can specify the distance between two consecutive labels applied on a duct.

The **Style**, **Layout method**, **Number of Labels**, and **Distance between** options are also available for Flow Arrows. These options work for flow arrows in the same way as for Labels.

# Flex Duct

This tool is used to create flexible ducts in the HVAC system. To create a flexible duct, choose the **Flex Duct** tool from the **Duct** drop-down in the **Build** panel of the **Home** tab in the **Ribbon**; the **Properties** palette will be displayed, refer to Figure 4-13. Some of the options available in the **Properties** palette after selecting flex duct have already been discussed in the Duct section of this chapter. Rest of the options are discussed next.

## Segment

This drop-down list is available in the **BASIC > Routing** rollout in the **Properties** palette. There are three options available in this drop-down list: **Line**, **Arc**, and **Spline**. Using the **Line** option, you can create a flex duct in the form of a straight line. Using the **Arc** option, you can create a flex duct in the form of an arc. Using the **Spline** option, you can create a flex duct in the form of a spline.

## Radius factor

This edit box is available in the **BASIC > Routing** rollout in the **Properties** palette. Using this edit box, you can specify the value of radius on the bend. The value entered in this edit box is multiplied by the diameter value of the flex duct gives the radius at the bend. This edit box is available only when the **Line** option is selected in the **Segment** drop-down list.

## Graphics

This drop-down list is available in the **ADVANCED > Graphics** rollout in the **Properties** palette. This drop-down list is available for both 1-Line and 2-Line annotations in the respective rollouts. Using the options available in this drop-down list, you can specify the shape and pattern of the annotative representation of the duct in the 2-D drawing.

## Pitch

This edit box is available in the **ADVANCED > Graphics** rollout in the **Properties** palette. This edit box is available for both 1-Line and 2-Line annotations in the respective rollouts. This edit box is used to specify the length of one piece flex duct.

## Vertex

This edit box is available in the **ADVANCED > Geometry** rollout in the **Properties** palette when you select an already created flex duct. Using this edit box, you can select an available vertex of the flexible duct. You can also change the selected vertex by using the scroll buttons available in the right of the edit box. On doing so, the options relevant to the selected vertex will be displayed.

## Vertex X

This edit box is available in the **ADVANCED > Geometry** rollout in the **Properties** palette. Using this edit box, you can specify X coordinate of the selected vertex.

## Vertex Y

This edit box is available in the **ADVANCED > Geometry** rollout in the **Properties** palette. Using this edit box, you can specify Y coordinate of the selected vertex

## Vertex Z

This edit box is available in the **ADVANCED > Geometry** rollout in the **Properties** palette. Using this edit box, you can specify Z coordinate of the selected vertex

# DUCT FITTING

Duct fitting is used to connect multiple ducts. The tools to add duct fitting are available in the **Duct Fitting** drop-down in the **Build** panel of the **Home** tab in the **Ribbon**. There are three tools available in this drop-down; **Duct Fitting**, **Duct Custom Fitting**, and **Duct Transition Utility**. These tools are discussed next.

## Duct Fitting

This tool is used to add a user specified fitting in the duct. On choosing this tool, the **Properties** palette is displayed, refer to Figure 4-18. Various options available in this palette are discussed next.

*Figure 4-18 The **Properties** palette displayed after choosing the **Duct Fitting** tool*

### Description

This field is available in the **BASIC > General** rollout in the **Properties** palette. When you click in this field, the **Description** dialog box will be displayed. You can write description about the fitting in this text box.

### System

This drop-down list is available in the **BASIC > General** rollout of the **Properties** palette. There are various options available in this drop-down list to change the standard of fitting to be created. By default, the **Standard (STD)** option is selected in this drop-down list.

### Part

This field is available in the **BASIC > General** rollout of the **Properties** palette. When you click in this field, the **Select a part** dialog box will be displayed, refer to Figure 4-19. There are various types of fittings available for duct in this dialog box. All these fittings are displayed in the **SELECT PART FROM CATALOG** rollout. The fittings in this rollout are divided into groups in a tree structure. You can filter the available parts depending on your requirements by using the options in the **FILTER** rollout of this dialog box. The filtered options will be displayed in the **SELECT PART SIZE** area of this dialog box. You can also select the required fitting from the **HVAC Tool** palette. These fittings are available in the **Fitting** tab of the palette.

*Figure 4-19  The Select a part dialog box*

### Current Size

This drop-down list is available in the **BASIC > General** rollout of the **Properties** palette. The options available in this drop-down list are used to change the size of the current selected fitting.

### Connections Rollout

This rollout is available in the **BASIC > Dimensions** rollout in the **Properties** palette. The options in this rollout are used to change the dimensions of a selected fitting. The options in this rollout may vary depending on the selected fitting.

### Elevation

This edit box is available in the **BASIC > Placement** rollout in the **Properties** palette. Using this edit box, you can specify the value of height of fitting from the ground level.

### Specify rotation on screen

This drop-down list is available in the **BASIC > Placement** rollout in the **Properties** palette. There are two options available in this drop-down list; **Yes** and **No**. If you select the **Yes** option, you will be prompted to specify the rotation value on screen while creating the fitting. If you select the **No** option, then you need to specify the rotation value in the edit box available in the **Properties** palette.

## Rotation

This edit box is available in the **BASIC > Placement** rollout in the **Properties** palette. Using this edit box, you can specify the value of rotation of the fitting before placing it in the drawing. This edit box is available only when the **No** option is selected in the **Specify rotation on screen** drop-down list.

## Vanes

This drop-down list is available in the **ADVANCED > Properties** palette. There are two options available in this drop-down list: **Yes** and **No**. If you select the **Yes** option, the vanes will be added to the fitting and if you select the **No** option, the vanes will not be added to the fitting.

## Duct Custom Fitting

Duct Custom Fitting    This tool is used for creating custom fittings in the duct line. To create a custom fitting, you must have lines, arcs, or polylines created in the drawing area. Choose the **Duct Custom Fitting** tool from the **Duct Fitting** drop-down; you will be prompted to select lines, arcs, or polylines. Select lines, arcs, or polylines from the drawing area and then press ENTER; the **Create Duct Custom Fitting** dialog box will be displayed, refer to Figure 4-20.

*Figure 4-20 The **Create Duct Custom Fitting** dialog box*

Depending on the entities selected, the segments will be displayed in the dialog box as Segment *n* of *m*, where *n* is the serial number of the current segment and *m* is the total number of segments in the fitting. Using the **Shape** drop-down list available in the **Segment *n* of *m*** area, you can change the shape of segments in the fitting. There are three options available in this drop-down list: **Rectangular**, **Round**, and **Oval**. After selecting the required option, you can specify the relevant parameters in the edit boxes available below this drop-down list. After specifying the parameters, You can choose the **Next >** button to change the parameters of next segment. You can also select the **Apply to All Segments** check box, if you want to keep same parameters for the other segments. After specifying parameters for all the segments, choose the **Finish** button; you will be prompted whether to delete the base geometry or not. Enter **Yes** or **No** at the command prompt; the custom fitting will be created.

## Duct Transition Utility

Duct Transition Utility     This tool is used to connect two parallel ducts, duct fittings, or equipment. Various shapes that can be connected by using these tools are given next.

| | | |
|---|---|---|
| Round | to | Round |
| Round | to | Rectangular |
| Round | to | Oval |
| Rectangular | to | Rectangular |
| Rectangular | to | Oval |
| Rectangular | to | Round |
| Oval | to | Oval |
| Oval | to | Rectangular |
| Oval | to | Round |

To create a duct transition, choose the **Duct Transition Utility** tool from the **Duct Fitting** drop-down; you will be prompted to select the first duct, duct fitting, or MvPart. Select the first entity; you will be prompted to select the second entity. Select the duct, duct fitting, or MvPart to be joined; a transition fitting will be created, joining both the entities.

# TUTORIAL

## Tutorial 1

In this tutorial, you will add a mechanical system (HVAC system) in the building created in Tutorial 2 of Chapter 3. You can also download the architectural file from *www.cadcim.com*. Path of the file is: *Textbooks > CAD/CAM > AutoCAD MEP > AutoCAD MEP 2014 for Designers > Tutorial*. The building after adding the mechanical system is shown in Figure 4-21. The plan for the mechanical system is given in Figure 4-22.

**(Expected time: 30 min)**

*Figure 4-21 The drawing after adding mechanical system*

*Figure 4-22* *The plan for mechanical system*

The following steps are required to complete this tutorial:

a. Calculate the zone wise airflow.
b. Install the Air Terminals on the basis of airflow required in a zone.
c. Add the duct line to the Air Terminals
d. Install the AHU on the basis of requirement.
e. Calculate the size of duct on the basis of requirements.
f. Install the duct lines and duct fittings in the system.

Now, you need to open the project file created in Tutorial 2 of Chapter 3.

## Opening the Drawing File

1. Choose **Open > Drawing** from the **Application** menu; the **Select File** dialog box is displayed.

2. Select the file created in Tutorial 2 of Chapter 3 and choose the **Open** button, the file is opened.

## Calculating Zone-wise Airflow

Before calculating the airflow, you need to divide the HVAC covered area into zones, refer to Figure 4-23.

1. Find out the volume of each zone on the basis of area and height of the zones.

2. On the basis of the volume of zones, the total air flow required can be given as:

Air Flow for Zone1 = 144 m3/h = 40 l/s

Air Flow for Zone2 = 180 m3/h = 50 l/s

Air Flow for Zone3 = 720 m$^3$/h = 200 l/s

*Figure 4-23* *The drawing after dividing into sections*

## Installing Air Terminals

Depending on the air flow required for each zone and the area to be covered, you need to install the diffusers.

1.   Choose the **Air Terminal** tool from the **Equipment** drop-down in the **Build** panel of the **Home** tab in the **Ribbon**; the **Add Multi-view Parts** dialog box is displayed with the options related to diffusers.

2.   Select **600 x 600 mm Square Faced Ceiling Diffuser** from the part list displayed in the **Part** tab of the dialog box; the parts for the selected category are displayed in the **Part Size Name** drop-down list.

3.   Select the part with 150mm neck from the drop-down list, set the value of elevation as **4500**, and install the diffusers according to the air flow requirements in the zones, refer to Figure 4-24.

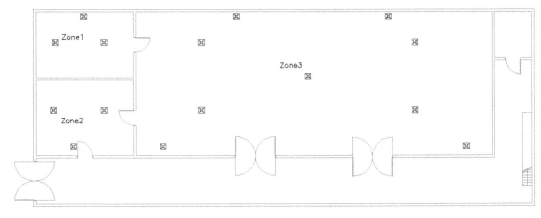

*Figure 4-24* *The drawing after adding diffusers*

4. Select the diffuser in the center of Zone1 and Zone2. Change the Flow rate as 20 l/s in the **Extended Data** tab of the **Properties** palette.

5. Select the five diffusers in the center of Zone3 and change the Flow rate as 40 l/s in the **Extended Data** tab of the **Properties** palette.

6. Select the two diffusers adjacent to the wall of Zone1 and Zone2. Change the Flow rate as 40 l/s in the **Extended Data** tab of the **Properties** palette.

7. Select the four diffusers adjacent to the wall of Zone3 and change the Flow rate as 50 l/s in the **Extended Data** tab of the **Properties** palette.

## Adding the Duct Line to the Diffusers

Now, you need to route the duct line by using the **1-Line** tool.

1. To create routing of the duct line, choose the **1-Line** tool from the **By Shape** area of the **Duct** tab in the Tool Palette; you will be prompted to specify the starting point of the duct line.

2. Select the **Fresh Air - Make-up (FA MAKE-UP)** option from the **System** drop-down list in the **BASIC > General** rollout of the **Design** tab in the **Properties** palette and create the fresh air duct line, as shown in Figure 4-25.

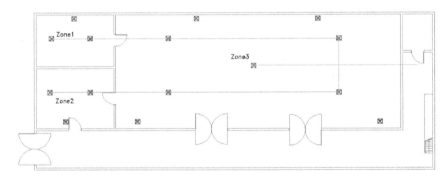

*Figure 4-25  The drawing after creating fresh air duct line*

3. Again, choose the **1-Line** tool and select the **Return (RA)** option from the **System** drop-down list in the **General** rollout of the **BASIC** rollout of the **Design** tab in the **Properties** palette.

4. Create the return duct line, as shown in Figure 4-26.

## Adding an Air Handler

Now, you need to add an Air Handler in the drawing to complete the duct line.

1. Choose the **Air Handler** tool from the **Equipment** drop-down in **Build** panel of the **Home** tab in the **Ribbon**; the **Add Multi-view Parts** dialog box is displayed, as shown in Figure 4-27.

*Figure 4-26* *The drawing after creating the return duct line*

*Figure 4-27* *The **Add Multi-view Parts** dialog box*

2. Select the **Air Handling Units - Floor Mounted Front Discharge** equipment from the Parts list in the **Part** tab of this dialog box. Preview of the equipment is displayed on the right of the Part list.

3. Specify the elevation value as **4500** in the **Elevation** edit box.

4. Click in the drawing area to specify the location of equipment; you are prompted to specify the rotation angle for equipment.

5. Enter the angle value as **0**; the equipment is placed at the specified location, refer to Figure 4-28.

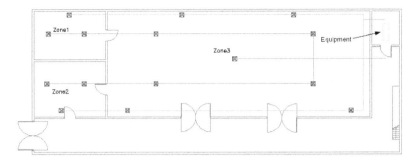

*Figure 4-28* *The drawing after adding equipment*

Now, you need to add duct line to the equipment.

6.   Select the equipment; the **+** sign will be displayed on various ports of the equipment.

7.   Click on the **+** signs one by one and connect the equipment, as shown in Figure 4-29.

*Figure 4-29* *The drawing after connecting the equipment with the duct line*

## Calculating Duct Size and Installation
Now, you need to calculate the duct size of each segment.

1.   Select the duct lines, as shown in Figure 4-30 and choose the **Calculate Duct Sizes** button
     from the **Calculations** panel of the **Duct** contextual tab in the **Ribbon**; the **Duct System
     Size Calculator** dialog box is displayed, refer to Figure 4-31.

2.   Change all the settings in the dialog box according to Figure 4-31.

3.   Choose the **Start** button from the **4** area of the dialog box; the **Multiple Parts Found**
     dialog box is displayed, refer to Figure 4-32.

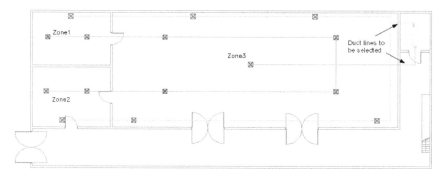

*Figure 4-30* *The duct lines to be selected for calculating duct size*

*Figure 4-31* *The **Duct System Size Calculator** dialog box*

4.  Select the desired part from the **Select a detailed part** area of the dialog box and choose the **OK** button; the selected part is added at the prompted position.

5.  Similarly, add all the parts at the prompted locations one after the other.

6.  After adding all the parts, the **Duct System Size Calculator** dialog box is displayed, as shown in Figure 4-33.

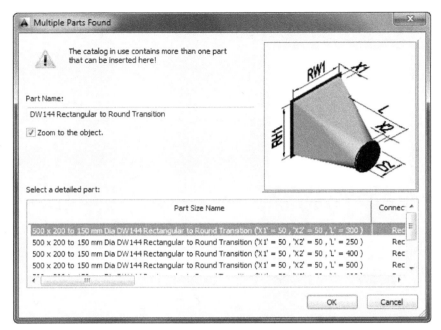

*Figure 4-32  The **Multiple Parts Found** dialog box*

7.  If the success rate is displayed as 100%, choose the **Close** button and exit the dialog box.

8.  If success rate is not displayed as 100%, choose the **View Event Log** button from the **5** area of the dialog box; the **Event Log** dialog box is displayed, refer to Figure 4-34.

9.  Select the object in the **Object** column that shows an error in the **Error** column, refer to Figure 4-34, and then choose the **Highlight and Zoom the selected object** button from the dialog box; the selected object is highlighted in the drawing area, refer to Figure 4-35.

10. Rectify the errors according to the description given in the **Description** column of the **Event Log** dialog box and then choose the **Start** button again after all the errors are rectified.

11. The drawing after calculating the duct size is displayed, as shown in Figure 4-36.

***Figure 4-33*** *The **Duct System Size Calculator** dialog box after completing calculations*

***Figure 4-34*** *The **Event Log** dialog box*

*Figure 4-35*  *The object causing error highlighted in the drawing area*

*Figure 4-36*  *The drawing after calculating duct size*

## Saving the drawing file

1.  Choose **Save** from the **Application** menu to save the drawing file.

## Self-Evaluation Test

**Answer the following questions and then compare them to those given at the end of this chapter:**

1.  The **Air Handler** option is available in the **Equipment** drop-down of the **Build** panel. (T/F)

2. The AHU Inspection module is used to check the quantity of air to be spread in the desired area. (T/F)

3. An AHU Mixing Box has three ports, one for inlet of _____, one for inlet of fresh air, and one for _____ of mixed air.

4. You can specify the flow rate of an air terminal by using the_____ edit box.

5. In which of the following drop-down is the **Fan** tool available?

   (a) **Equipment**                    (b) **Duct Fitting**
   (c) **Duct**                         (d) **Pipe**

## Review Questions

**Answer the following questions:**

1. The AHU Mixing Box is an equipment used to vary the speed of air flowing through the duct. (T/F)

2. Using the **Elevation** edit box, you can specify the value of elevation of the object from ground. (T/F)

3. There are three options available in AutoCAD MEP for specifying bends of a duct; _____, _____, and _____.

4. For branch fitting, you can use two options, _____ and _____.

5. Which of the following buttons is used to calculate the duct size of a duct system?

   (a) **Calculate Duct Sizes**         (b) **Duct**
   (c) **Flexible Duct**                (d) **Detail Components**

# EXERCISE

## Exercise 1

In this exercise, you will add an HVAC system in an office. The office after adding the HVAC system is shown in Figure 4-37. The plan for the HVAC system is given in Figure 4-38. You can download the architecture model of the office from *www.cadcim.com*.

**(Expected time: 30 min)**

*Figure 4-37* The HVAC system in the office

*Figure 4-38* The plan for the HVAC system

## Answers to Self-Evaluation Test
**1.** T, **2.** F, **3.** returning air, outlet, **4. Flow (Each Terminal)**, **5. Equipment**

# Chapter 5

# Creating Piping Systems

## Learning Objectives

**After completing this chapter, you will be able to:**

- *Use equipment required in a piping system*
- *Change basic settings of a piping system*
- *Configure pipe line options*
- *Route the pipe line*
- *Create pipe and flex pipe*
- *Create a custom part*

# INTRODUCTION

In this chapter, you will learn the use of various equipment for creating a piping system. A piping system is used to supply the desired amount of water to a specific location. A piping system is composed of piping equipment and pipes. Piping equipment and pipes together control the temperature, pressure and flow rate of the fluid passing through the pipes. To create a piping system, you need to know the vertical heads, pressure, temperature, and flow rate of the fluid. These variables are used for deciding the capacity of the equipment that are to be added in the piping system. To create a piping system, choose the **Piping** option from the **Workspace Switching** flyout; the **Piping** workspace will be invoked. The equipment available in the **Piping** workspace of AutoCAD MEP are discussed next.

# ADDING EQUIPMENT

For creating a piping system, you need to add related equipment to the given structure. All the equipment that can be added while working in the **Piping** workspace are available in the **Equipment** drop-down of the **Build** panel in the **Home** tab of the **Ribbon**, refer to Figure 5-1. The tools available in this drop-down are discussed next.

## Heat Exchanger

A heat exchanger is an equipment or a device which is used to transfer heat efficiently between two fluids. This heat transfer can occur through conduction or convection. To add a heat exchanger to the drawing, choose the **Heat Exchanger** tool from the **Equipment** drop-down available in the **Build** panel of the **Home** tab in the **Ribbon**; the **Add Multi-view Parts** dialog box will be displayed, as shown in Figure 5-2. Select the **Shell and Tube Heat Exchangers** is selected in the **Heat Exchanger** category of the **Mechanical** node. The preview of the selected component is displayed in the right of the dialog box. The D1, D2, D3, and D4 in the Preview area indicate the diameters of ports available on the heat exchanger. Additionally, you can select the required size for the

*Figure 5-1  The Equipment drop-down*

heat exchanger from the **Part Size Name** drop-down list. You can specify the elevation value of heat exchanger from the ground or UCS by using the **Elevation** edit box available below the **Part Size Name** drop-down list. After specifying the required parameters, click in the drawing area to place the heat exchanger; a compass will be displayed below the heat exchanger. Using this compass, you can rotate the heat exchanger at the required angle. After specifying the angle, choose the **Close** button from the **Add Multi-view Parts** dialog box to exit it. Figure 5-3 shows a Shell and Tube Heat Exchanger.

*Figure 5-2* The **Add Multi-view Parts** *dialog box*

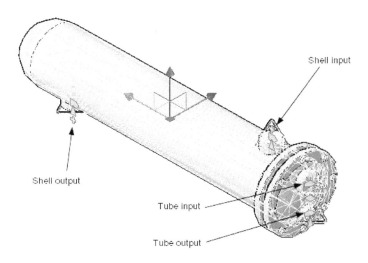

*Figure 5-3* The Shell and Tube Heat Exchanger

# Pump

A pump is a mechanical device that is used to supply fluid to the desired point by using mechanical force. To add a pump to the drawing, choose the **Pump** tool from the **Equipment** drop-down available in the **Build** panel of the **Home** tab in the **Ribbon**; the **Add Multi-view Parts** dialog box will be displayed, as shown in Figure 5-4. Select the **Base Mounted Pumps** sub-category in the **Pumps** category of **Mechanical** node. Select the required sub-category from the dialog box; various sizes available for that sub-category are displayed in the **Part Size Name** drop-down list. Select the required size from the drop-down list; you will be prompted to specify a point in the drawing area to place the pump. Click in the drawing area to place the pump; a compass will be displayed below the pump. Using this compass, you can rotate the pump at the required angle. After specifying the angle, choose the **Close** button from the **Add Multi-view Parts** dialog box to exit it. Figure 5-5 shows a base mounted pump.

*Figure 5-4  The **Add Multi-view Parts** dialog box displayed after selecting the **Pump** tool*

*Figure 5-5  A base mounted pump*

## Tank

A tank is a hollow space created in a structure to store the fluid for later use. There are two types of tanks available in AutoCAD MEP: Expansion Tank and Storage Tank. The expansion tank acts as a safety tank for fluid heating system. This tank allows the fluid to expand so that it can release pressure. The storage tank is used to store the fluid for later use. To add a tank to the drawing, choose the **Tank** tool from the **Equipment** drop-down available in the **Build** panel of the **Home** tab in the **Ribbon**; the **Add Multi-view Parts** dialog box will be displayed, as shown in Figure 5-6. Select the **Freestanding Vertical Expansion Tanks** part from the **Expansion Tanks** sub-category of the **Tanks** category in Mechanical equipment. Select the required part from the dialog box; various sizes available for that sub-category are displayed in the **Part Size Name** drop-down list. Select the required size from the drop-down list and double-click in the drawing area to place the tank; a compass will be displayed below the component. Using this compass, you can rotate the component at the required angle. After specifying the angle, choose the **Close** button from the **Add Multi-view Parts** dialog box to exit the dialog box. Figure 5-7 shows the tanks available in AutoCAD MEP.

*Figure 5-6  The **Add Multi-view Parts** dialog box displayed after choosing the **Tank** tool*

## Valve

A valve is a device that is used to regulate or direct the flow of fluid through a passage. There are various types of valves available in AutoCAD MEP. These valves are available in two main categories: Fire Protection and Valves. To add a valve to the drawing, choose the **Valve** tool from the **Equipment** drop-down available in the **Build** panel of the **Home** tab in the **Ribbon**; the **Add Multi-view Parts** dialog box will be displayed, as shown in Figure 5-8. Select the **Gate Valves Hose Connection** part from the **Hose Connections** sub-category of the **Fire Protection** category in the **Mechanical** node. Select the required part from the dialog box; various sizes available for that sub-category are displayed in the **Part Size Name** drop-down list. Select the required size from the drop-down list and click in the drawing area to place the valve; a compass will be displayed below the valve. Using this compass, you can rotate the valve at the required angle. After specifying the angle, choose the **Close** button from the **Add Multi-view Parts** dialog box to exit it.

*Figure 5-7  The tanks available in AutoCAD MEP*

*Figure 5-8  The **Add Multi-view Parts** dialog box displayed after selecting the **Valve** tool*

## Equipment

When you choose the **Equipment** tool from the **Equipment** drop-down of the **Build** panel of the **Home** tab, the **Add Multi-view Parts** dialog box will be displayed, as shown in Figure 5-9. Click on the **+** sign adjacent to the **Mechanical** node in the **Part** tab; the mechanical equipment available in AutoCAD MEP will be displayed in a tree structure.

*Figure 5-9* *The **Add Multi-view Parts** dialog box*

Click on the **+** sign adjacent to the required component; various options related to the selected component will be displayed in a list. Select the required option from the list and click in the drawing area to place the component. A compass will also be displayed below the component. Using this compass, you can rotate the component at the required angle. After specifying the angle, choose the **Close** button from the **Add Multi-view Parts** dialog box to exit.

# CREATING PIPE LINES

In a piping system, pipe is a mechanical object through which the fluid flows. There are two types of pipes available in AutoCAD MEP: Pipe and Parallel Pipe. The tools to create these pipes are available in the **Pipe** drop-down in the **Build** panel of the **Home** tab in the **Ribbon**. These tools are discussed next.

## Pipe

This tool is used to create a pipe line in the piping system. The shape of the pipe created by using this tool is round. When you choose this tool, the **Properties** palette will be displayed, as shown in Figure 5-10. Also, you will be prompted to specify the start point of the pipe line. Click in the drawing area to specify the start point of the pipe line; you will be prompted to specify the next point of the pipe line. Specify the next point to create the pipe line. The options available in the **Properties** palette are discussed next.

### Description

This field is available in the **BASIC > General** rollout in the **Properties** palette. When you click in this field, the **Description** dialog box will be displayed, as shown in Figure 5-11. You can enter description for the pipe in the **Edit the description for this object** text box in this dialog box. Choose **OK** to exit the dialog box.

*Figure 5-10  The **Properties** palette on choosing the **Pipe** tool*

## System
This drop-down list is available in the **BASIC > General** rollout of the **Properties** palette. This drop-down list contains various system definitions for the pipe.

## Routing Preferences
This drop-down list is available in the **BASIC > Dimensions** rollout. The options available in this drop-down list are used for routing style of the pipe. By default, the **Standard** option is selected in this drop-down list.

## Nominal size
This drop-down list is available in the **BASIC > Dimensions** rollout. This drop-down list contains options to change the size of pipe. The options available in this drop-down list change depending upon the option selected in the **Routing preferences** drop-down list.

*Figure 5-11  The **Description** dialog box*

## Specify Cut Length

This drop-down list is available in the **General > Dimensions** rollout. There are two options available in this drop-down list: **Yes** and **No**. By default, the **No** option is selected in this drop-down list. On selecting the **Yes** option, you can specify the cut length in the respective field.

## Cut Length

This edit box is available in the **Dimensions** rollout only when the **Yes** option is selected in the **Specify Cut Length** drop-down list. You can specify the value of cut length in this edit box. Cut length is the maximum available length of a piece of pipe.

## Elevation

This edit box is available in the **BASIC > Placement** rollout. Using this edit box, you can specify the value of elevation of pipe from the ground.

## Start elevation

This field is available in the **BASIC > Placement** rollout. The value in this field is the same as specified in the **Elevation** edit box.

## Justify

This drop-down list is available in the **Placement > Justification** rollout. The options in this drop-down list are used to change the justification of pipe.

## Horizontal offset

This edit box is available in the **Justification** rollout of the **Placement** rollout. Using this edit box, you can specify the horizontal distance value by which the pipe is distant from the selected justification point.

## Vertical offset

This edit box is available in the **Justification** rollout of the **Placement** rollout. Using this edit box, you can specify the vertical distance value by which the pipe is distant from the selected justification point.

## Slope format

The options in this drop-down list are used to specify the format in which the slope value will be entered. There are seven options available in this drop-down list: **Angle (Decimal Degrees)**; **Percentage, 100%=45 degree**; **Percentage, 100%=90 degree**; **Rise/Run (Millimeters/Millimeters)**; **Run (Millimeters), Rise=(1 Millimeters)**; **Rise (Millimeters), Run= (10 Millimeters)**; and **Fractional Rise (Millimeters), Run= (10 Millimeters)**. In AutoCAD MEP 2014, the slope also includes the gravity factor in it.

## Slope

This edit box is available in the **BASIC > Routing** rollout of the **Properties** palette. You can specify the value of slope in this edit box. The format in which the value will be specified depends upon the option selected from the **Slope format** drop-down list. If you change the slope value while adding more pipes, the fitting at joints will adjust automatically to provide the changed slope.

## Bend method

This drop-down list is used to specify the method of bending that is to be applied on any bend while creating a pipe line. There are two methods of bending available in this drop-down list: **Elbow** and **Offset**.

## Bend angle

This drop-down list is used to specify the angle of bend that is to be applied on any bend while creating a pipe. By default, the **Value from Catalog** option is selected in this drop-down list. Therefore, the bend angle specified in the catalog is used. You can specify a desired value by clicking on this drop-down list.

## Branch angle

This drop-down list is used to specify the lateral angle of **Tee** that is to be applied on any bend while creating a pipe. By default, the **Value from Catalog** option is selected in this drop-down list. Therefore, the bend angle specified in the catalog is used. You can specify a desired value by clicking on this drop-down list.

## Connection details

This button is available in the **ADVANCED** rollout. When you choose this button, the **Connection Details** dialog box will be displayed, refer to Figure 5-12. There are two rollouts available in the dialog box which contain the details about both the connections.

*Figure 5-12*  The **Connection Details** dialog box

## Insulation thickness

This edit box is also available in the **ADVANCED** rollout and is used to specify the thickness of insulation applied in the pipe.

## Use fitting tolerance

This drop-down list is available in the **ADVANCED > Routing Options** rollout of the **Properties** palette. There are two options in this drop-down list: **Yes** and **No**. When the **Yes** option is selected in the drop-down list, the branch fittings will be connected to the main pipe, considering the fitting tolerances. If the **No** option is selected in the drop-down list, the branches will be joined to the main pipe without considering the fitting tolerances.

## Joint direction

This drop-down list is available in the **ADVANCED > Routing Options** rollout of the **Properties** palette. There are two options available in this drop-down list: **Male into Female** and **Female out to Male**.

## Branch fitting

This drop-down list is available in the **ADVANCED > Routing Options** rollout of the **Properties** palette. The options in this drop-down list are used to specify the type of fitting to be applied on a branch. There are five options available in this drop-down list: **Takeoff only**, **Tee only**, **Tee or Wye (Lateral)**, **Wye (Lateral) only**, and **Wye (Lateral) or Tee**.

## Fitting Setting

This option is available in the **ADVANCED > Routing Options** rollout of the **Properties** palette. When you click on this option, the **Fitting Settings** dialog box will be displayed, as shown in Figure 5-13. The options in this dialog box are used to change the type of fitting to be used while creating the branches in the pipe.

*Figure 5-13* The *Fitting Settings* dialog box

## Preferences

This option is available in the **ADVANCED > Routing Options** rollout of the **Properties** palette. When you click on this option, the **Pipe Layout Preferences** dialog box will be displayed, as shown in Figure 5-14. The options in this dialog box are used to define the layout preferences for creating a pipe. Using the options in this dialog box, you can set the preferences for parameters such as Slope, Elevation, Label style, Flow arrow style, and Elbow layout.

## Style

This drop-down list is available in the **ADVANCED > Labels and Flow Arrows > Labels** rollout of the **Properties** palette. The options in this drop-down list are used to apply the label styles to pipe.

*Figure 5-14  The **Pipe Layout Preferences** dialog box*

## Layout method

This drop-down list is available in the **ADVANCED > Labels and Flow Arrows > Labels** rollout of the **Properties** palette only when **NONE** is not selected from the **Style** drop-down list. The options in this drop-down list are used to specify the type of layout for pipes. There are two options available in this drop-down list: **By distance** and **By quantity**.

## Number of Labels

This edit box is available in the **ADVANCED > Labels and Flow Arrows > Labels** rollout of the **Properties** palette only when **NONE** is not selected in the **Style** drop-down list and **By quantity** is selected in the **Layout method** drop-down list. Using this edit box, you can specify the total number of labels on a segment pipe.

## Distance between

This edit box is available in the **ADVANCED > Labels and Flow Arrows > Labels** rollout of the **Properties** palette only when **NONE** is not selected in the **Style** drop-down list and **By distance** is selected in the **Layout method** drop-down list. Using this edit box, you can specify the distance between two consecutive labels on a pipe.

The **Style**, **Layout method**, **Number of Labels**, and **Distance between** options are also available for Flow Arrows. These options have already been discussed in this chapter.

## Parallel Pipes

⊟ Parallel Pipes  This tool is used to create parallel pipes in the Piping system. To create a parallel pipe, choose the **Parallel Pipes** tool from the **Pipe** drop-down in the **Build** panel of the **Home** tab in the **Ribbon**; you will be prompted to select a baseline object. Select a base line object, like an earlier created pipe, from the drawing area; you will be prompted to select parallel pipes. Select parallel pipes from the drawing area and press ENTER; you will be prompted to specify the next point for creating parallel pipes. The options available in the **Properties** palette for parallel pipes are same as discussed in the Pipe section of this chapter.

# ADDING PIPE FITTINGS

A pipe fitting is used to join two or more pipes. You can add pipe fittings by using the tools available in the **Pipe Fitting** drop-down. This drop-down is available in the **Build** panel of the **Home** tab in the **Ribbon**. The tools in this drop-down are discussed next.

## Pipe Fitting

⌐ Pipe Fitting  This tool is used to add a user specified fitting in the pipe line. On choosing this tool, the **Properties** palette will be displayed, refer to Figure 5-15. The options available in this palette are discussed next.

### Description

This field is available in the **BASIC > General** rollout in the **Properties** palette. When you click in this field, the **Description** dialog box will be displayed. You can write description about the fitting in the text box available in this dialog box.

### System

This drop-down list is available in the **BASIC > General** rollout of the **Properties** palette. The options in this drop-down list are used to change the standard of fitting to be created. By default, the **Standard (STD)** option is selected in this drop-down list.

### Part

This field is available in the **BASIC > General** rollout of the **Properties** palette. When you click in this field, the **Select a part** dialog box will be displayed, refer to Figure 5-16. There are various types of fittings available for pipe in this dialog box. All these fittings are displayed in the **SELECT PART FROM CATALOG** rollout. The fittings in this rollout are divided into groups in a tree structure. You can filter the available parts depending on your requirements by using the options available in the **FILTER** rollout of this dialog box. The filtered options will be displayed in the **SELECT PART SIZE** area of this dialog box. You can also select the required fitting from the **Piping Tool** palette. These fittings are available in the **Fitting** tab of the palette.

*Figure 5-15  The **Properties** palette for pipe fittings*

## Current size

This drop-down list is available in the **BASIC > General** rollout of the **Properties** palette. There are various options available in this drop-down list to change the size of the current fitting.

## Connections rollout

This rollout is available in the **BASIC > Dimensions** rollout in the **Properties** palette. There are various options available in this rollout to change the dimensions of a selected fitting. Options in this rollout vary depending on the selected fitting.

## Elevation

This edit box is available in the **BASIC > Placement** rollout in the **Properties** palette. Using this edit box, you can specify the value of height of fitting from the ground level.

## Specify rotation on screen

This drop-down list is available in the **BASIC > Placement** rollout in the **Properties** palette. There are two options available in this drop-down list: **Yes** and **No**. If you select the **Yes** option, then you will be prompted to specify the rotation value while creating the fitting. If you select the **No** option, then you need to specify the rotation value in the edit box available in the **Properties** palette.

*Figure 5-16  The Select a part dialog box*

## Rotation

This edit box is available in the **BASIC > Placement** rollout in the **Properties** palette. Using this edit box, you can specify the value of rotation of the fitting before placing it in the drawing. This edit box is available only when the **No** option is selected from the **Specify rotation on screen** drop-down list.

## Connection details

This button is available in the **ADVANCED** rollout. When you choose this button, the **Connection Details** dialog box will be displayed, as shown in Figure 5-17. There are two rollouts available in the dialog box that contain the details about both the connections.

## Insulation thickness

This edit box is also available in the **ADVANCED** rollout and is used to specify the thickness of insulation applied to the pipe.

# Pipe Custom Fitting

This tool is used for creating custom fittings in the pipe line. To create a pipe custom fitting, you must have lines, arcs, or polylines created in the drawing area. To create a custom fitting, choose the **Pipe Custom Fitting** tool from the **Pipe Fitting** drop-down list; you will be prompted to select lines, arcs, or polylines. Select lines, arcs, or polylines from the drawing area and press ENTER; the **Create Pipe Custom Fitting** dialog

*Figure 5-17  The **Connection Details** dialog box*

box will be displayed, refer to Figure 5-18. Depending on the selected entities, the segments will be displayed in the dialog box as *Segment n of m*, where *n* is the serial number of current segment and *m* is the total number of segments in the fitting. Using the **Shape** drop-down list available in the **Segment n of m** area, you can change the shape of segments in the fitting. There are various system definitions available in the **System** drop-down. After selecting the desired system, choose the **Next >** button to change the parameters of next segment. You can also select the **Apply to All Segments** check box if you want to keep the same parameters for other segments. After specifying parameters for all the segments, choose the **Finish** button; you will be prompted whether to delete the base geometry or not. Enter **Yes** or **No** at the Command prompt; the custom fitting will be created. The open end points of entities will be the connection ports for fitting.

*Figure 5-18  The **Create Pipe Custom Fitting** dialog box*

# CREATING A CUSTOM MULTI-VIEW PART

While creating models in AutoCAD MEP, you may need to create some custom parts that can be frequently used in the drawing. The procedure to create a custom Multi-view part is given next.

1. Create a model of the custom part that you want to add to the Multi-view Part list and save it as a block.

2. Choose the **Content Builder** tool from the **MEP Content** panel in the **Manage** tab of the **Ribbon**; the **Getting Started - Catalog Screen** dialog box will be displayed, as shown in Figure 5-19.

3. Select the **Multi-view Part** option from the **Part Domain** drop-down list available at the top of the dialog box; the options in the dialog box will be modified according to Multi-view parts.

4. Select **Mechanical** from the part tree displayed in the dialog box to add a new equipment range and choose the **New Chapter** button available at the right of the part tree; the **New Chapter** dialog box will be displayed, as shown in Figure 5-20.

5. Enter **Turbine** as name for the category in the **Name** edit box of the dialog box and choose **OK**; a new category will be added in the Mechanical part list. Figure 5-21 shows Turbine category added in the Mechanical part list.

*Figure 5-19* *The* *Getting Started - Catalog* *Screen* *dialog box*

*Figure 5-20* *The* *New Chapter* *dialog box*

*Figure 5-21  The **Getting Started - Catalog Screen**
dialog box with Turbine added as a new chapter*

6.  Choose the **New Block Part** button from the **Getting Started - Catalog Screen**
    dialog box; the **New Part** dialog box will be displayed, as shown in Figure 5-22.

*Figure 5-22  The **New Part** dialog box*

7.  Enter **Steam Turbine** and description of the new part in the **Name** and **Description** edit
    boxes, respectively and choose **OK**; the **MvPart Builder (New Part) - Steam Turbine**
    dialog box will be displayed, refer to Figure 5-23.
8.  Select an option from the **Type** drop-down list whose properties are matching with the
    part you want to create. In this case, select the **Cooling Tower** option from this drop-down
    list; the options in the **Layer Key** and **Subtype** fields will be selected accordingly.

**Figure 5-23** *The MvPart Builder (New Part) - Steam Turbine dialog box displayed*

9. Choose the **Next** button from the dialog box; the **Blocks & Names** page of the dialog box will be displayed, as shown in Figure 5-24.

**Figure 5-24** *The MvPart Builder (New Part) - Steam Turbine dialog box with the Block & Names page displayed*

10. Choose the **Add Part Size** button available below the part sheet; a new part will be  added in the list and you will be prompted to select an option from the drop-down list displayed.

11. Select the block you have created for the part one by one in the respective fields; the blocks will be assigned to the new part. (In this case the block is with the name turbine).

12. Choose the **Generate Blocks** button to create the part; the **Views** dialog box will be displayed, as shown in Figure 5-25.

*Figure 5-25  The **Views** dialog box*

13. Using the options available in this dialog box, you can change the display style of the part to be created. After configuring the display style, choose the **OK** button to exit.

14. Choose the **Next** button from the **MvPart Builder (New Part) - Steam Turbine** dialog box displayed; the **Image** page will be displayed, as shown in Figure 5-26.

15. Select the **Generate an image based on a model block from SW Isometric View** radio button and then choose the **Generate** button from this page; the image of block will be generated. After generating the image, choose the **Next** button; the **Connectors** page will be displayed, as shown in Figure 5-27.

16. Using the options available in this page, you can specify the connectors to be added to the Multi-view part. To add a connector to the part, right click on the part; a shortcut menu will be displayed, as shown in Figure 5-28.

*Figure 5-26*  *The* ***MvPart Builder (New Part) - Steam Turbine***
*dialog box with the* ***Image*** *page*

*Figure 5-27*  *The* ***MvPart Builder (New Part) - Steam Turbine***
*dialog box with the* ***Connectors*** *page*

Add Cable Tray Connector...

Add Conduit Connector...

Add Duct Connector...

Add Pipe Connector...

*Figure 5-28  Shortcut menu to add connectors*

17. Choose the desired option from the shortcut menu; the **Part Family Connector Properties** dialog box will be displayed. Figure 5-29 shows the dialog box displayed after choosing the **Add Pipe Connector** option.

*Figure 5-29  The **Part Family Connector Properties** dialog box*

18. Click on the field adjacent to **Flow Direction**; a drop-down list will be displayed. Select the **In** option from the drop-down list.
19. Similarly, add one more pipe connector with the **Out** option and choose the **OK** button from the dialog box; the **MvPart Builder (New Part)** dialog box after adding the connectors will be displayed, as shown in Figure 5-30.
20. Select and right-click on **Connector1** available below **Part size 1** in the list; a shortcut menu will be displayed.
21. Choose the **Edit Placement** option from the shortcut menu; the Application window will be displayed, as shown in Figure 5-31 and you will be prompted to specify the position or normal, of connector. Also, the **MvPartBuilder - Connector Editor** palette is displayed.

*Figure 5-30  The **MvPart Builder (New Part)** dialog box after adding connectors*

*Figure 5-31  The Application window*

22. Choose the **Position** button from the command prompt and specify the location of the first connector.

23. Similarly, specify the location of the second connector after selecting **Connector2** from the **MvPartBuilder - Connector Editor** palette.

24. After specifying the locations, choose the **OK** button from the **MvPartBuilder - Connector Editor** palette; the **MvPart Builder** dialog box will be displayed again.

25. Choose the **Next** button from the dialog box; the **Properties** page will be displayed, as shown in Figure 5-32.

*Figure 5-32* *The MvPart Builder dialog box with the* *Properties* *page displayed*

26. Choose the **Edit Properties** button from the dialog box; the **Property Editor** dialog box will be displayed, refer to Figure 5-33. Using the options available in this dialog box, you can add desired property variables to the new part.

27. After configuring the property variables, choose the **OK** button and then choose the **Finish** button from the **MvPart Builder** dialog box; the part will be created and added to the equipment list.

*Figure 5-33*  *The **Property Editor** dialog box*

# TUTORIAL

## TUTORIAL 1

In this tutorial, you will create the pipe system shown in Figure 5-34. The plan view of the pipe system is given in Figure 5-35. Note that to complete this piping system, you also need to create custom part named steam turbine component.          **(Expected time: 30 min)**

*Figure 5-34*  *The piping system*

Legends

C — 33.5 KW Large Horizontal Air Cooled
Condenser
T — custom steam turbine
B — 600 kW Water tube Boilers
p — 80x80 mm Base Mounted Pump

*Figure 5-35* *Plan of the piping system*

The following steps are required to complete this tutorial:

Examine the model to determine equipment to be added and parameters of piping.

a.   Create a custom part named Turbine.
b.   Add the equipment according to the model .
c.   Create piping between various equipment.

## Creating the Drawing File

1.   Choose **New > Drawing** from the **Application** menu; the **Select Template** dialog box is displayed.

2.   Select the **Aecb Model (Global Ctb).dwt** template from the dialog box and then choose the **Open** button; a blank drawing file is created.

## Creating a Custom Part

To create a custom part, you must have a model of the part created as a block. So, before creating a custom part, you need to create a model of the part and save it as a block.

Before creating a model, you need to add the **Solids** tab in the **Ribbon**. To do so, right-click anywhere on the **Ribbon**; a shortcut menu is displayed. Choose the **Solids** option from the **Show Tabs** flyout in the shortcut menu; the **Solids** tab is added in the **Ribbon**.

1.   Choose the **Cone** tool from the **Solid Primitives** drop-down in the **Modeling** panel of the **Solids** tab in the **Ribbon**; you will be prompted to specify the center of the base.

2. Enter **0,0,0** at the Command prompt and create a cone with the base radius **1000**, top radius **500**, and height **1000**. Also, create the inlet and outlet pipes, refer to Figure 5-34.

3. Select the cone and then choose the **Create Block** tool from the **Block** panel of the **Insert** tab in the **Ribbon**; the **Block Definition** dialog box is displayed, as shown in Figure 5-36.

*Figure 5-36  The **Block Definition** dialog box*

4. Specify the name of the block as **Steam Turbine** in the **Name** edit box available at the top left corner of the dialog box and choose the **OK** button from the dialog box; the model is added as a block.

## Creating Custom Part by Using the Block

1. Choose the **Content Builder** tool from the **MEP Content** panel in the **Manage** tab of the **Ribbon**; the **Getting Started-Catalog Screen** dialog box is displayed, as shown in Figure 5-37. Select the **Multi-view Part** option from the **Part Domain** drop-down list if not selected.

2. Click on **Mechanical** in the part tree and choose the **New Chapter** button at the right of the dialog box; the **New Chapter** dialog box is displayed, as shown in Figure 5-38.

3. Specify the name as **Turbine** in the **Name** edit box of the dialog box and then choose the **OK** button to exit; Turbine is added in the Mechanical part list.

4. Select **Turbine** from the list and choose the **New Block Part** button at the right of the dialog box; the **New Part** dialog box is displayed, as shown in Figure 5-39.

5.  Specify the name of the part as **Steam Turbine** in the **Name** edit box and then click in the **Description** edit box; the text specified in the **Name** edit box is copied in the **Description** edit box.

6.  Choose the **OK** button from the dialog box; the **MvPart Builder (New Part)** dialog box is displayed, refer to Figure 5-40.

**Figure 5-37** The **Getting Started- Catalog Screen** dialog box

**Figure 5-38** The **New Chapter** dialog box

**Figure 5-39** The **New Part** dialog box

*Figure 5-40  The MvPart Builder (New Part) dialog box*

7.  Select the **Pump** option from the **Type** drop-down list and select the **Inline Pump** option from the **Sub-Type** drop-down list.

8.  Choose the **Next** button from the dialog box; the **Blocks & Names** page of the dialog box is displayed, as shown in Figure 5-41.

9.  Choose the **Add Part Size** button available below the part size sheet; a drop-down list is displayed.

10. Select the **Steam Turbine** option from the drop-down list and then choose the **Generate Blocks** button from the **MvPart Builder (New Part)** dialog box; the **Views** dialog box is displayed, as shown in Figure 5-42.

11. Choose the **OK** button from the **Views** dialog box; the **MvPart Builder (New Part)** dialog box is displayed. Choose the **Next** button from it; the **Image** page of the dialog box is displayed, as shown in Figure 5-43.

**Figure 5-41** *The* **MvPart Builder (New Part)** *dialog box with the* **Blocks & Names** *page displayed*

**Figure 5-42** *The* **Views** *dialog box*

12. Select the **Generate an image based on a model block from the SW Isometric View** radio button, and then choose the **Generate** button; preview of the block is displayed in the Preview area of the dialog box.

13. Choose the **Next** button from the dialog box; the **Connectors** page is displayed, refer to Figure 5-44.

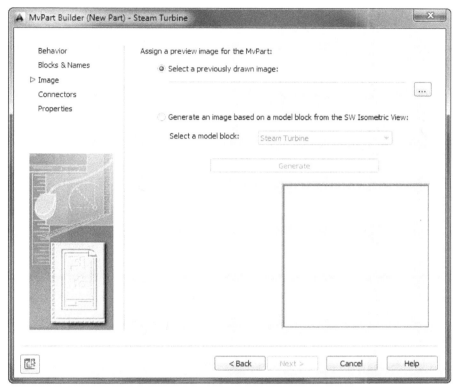

*Figure 5-43  The MvPart Builder (New Part) dialog box with the Image page*

14. Right-click on **Steam Turbine** with symbol, refer to Figure 5-44; a shortcut menu is displayed.

15. Choose the **Add Pipe Connector** option from the shortcut menu; the **Part Family Connector Properties** dialog box is displayed, as shown in Figure 5-45.

16. Select the **In** option from the **Flow Direction** drop-down list and the **Hot Water** option from the **System Type** drop-down list.

17. Choose the **OK** button to close the dialog box; **Connector 1** is added to the new part.

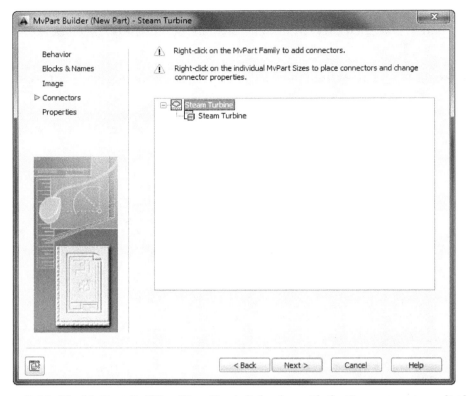

*Figure 5-44  The MvPart Builder (New Part) dialog box with the Connectors page displayed*

*Figure 5-45   The Part Family Connector Properties dialog box*

18. Similarly, add another pipe connector to the part and select the **Out** option from the **Flow Direction** drop-down list and the **Hot Water** option from the **System** drop-down list.

19. Right-click on the **Connector 1** available below **Steam Turbine** with symbol; a shortcut menu is displayed.

20. Choose the **Edit Placement** option from the shortcut menu; the **MvPartBuilder - Connector Editor** palette is displayed, as shown in Figure 5-46 and you are prompted to specify position of connector or normal to connector.

21. Choose the **Position** button from the Command prompt; you are prompted to select position for the first connector. Click on the position displayed as In Connector in Figure 5-47 and select the **Flange** option from the **Connection Type** drop-down list in the **MvPartBuilder - Connector Editor** palette. Click in the drawing area and choose the **OK** button to accept the results.

22. Similarly, specify the position of Connector 2 and then choose the **OK** button from the **MvPartBuilder - Connector Editor** palette; the **MvPart Builder (New Part)** dialog box is displayed again.

23. Choose the **Next** button from the dialog box; the **Properties** page of the dialog box is displayed, as shown in Figure 5-48.

24. Choose the **Finish** button; the custom part is created with the name **Steam Turbine** in the **Turbine** category of equipment in the equipment list.

*Figure 5-46* *The MvPartBuilder - Connector Editor palette*

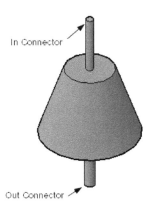

*Figure 5-47* *The model with In Connector and Out Connector*

*Figure 5-48* The *MvPart Builder (New Part)* dialog box with the *Properties* page displayed

## Adding the Equipment

1. Choose the **Equipment** tool from the **Equipment** drop-down available in the **Build** panel of the **Home** tab in the **Ribbon**; the **Add Multi-view Parts** dialog box is displayed.

2. Expand the Mechanical category; the list of all the available mechanical equipment is displayed.

3. Click on the **+** sign adjacent to Boilers; various types of boilers are displayed.

4. Select Firetube Boilers from the list, select the **600 kW Firetube Boilers** option from the **Part Size Name** drop-down list, and click anywhere in the drawing area to place the boiler. Place it with zero rotation.

5.  Similarly, add 80x80mm Base Mounted Pump, 33.5 KW Large Horizontal Air-Cooled Condenser, and the Steam Turbine created earlier, as shown in Figure 5-49 . For position, refer to Figure 5-35.

*Figure 5-49  The drawing after adding equipment list*

6.  Choose the **Close** button to exit the dialog box.

 **Note**
*Steam Turbine is available in the **Turbine** category of equipment.*

## Creating Piping between Various Equipment

1.  Choose the **Pipe** tool from the **Pipe** drop-down in the **Build** panel of the **Home** tab in the **Ribbon**; you are prompted to specify the start point of the pipe.

2.  Click on the Pipe End Connector available on the pump, refer to Figure 5-50, move the cursor in the vertical direction, and enter **1000** at the dynamic prompt displayed; the **Choose a Part** dialog box is displayed, refer to Figure 5-51.

*Figure 5-50  The Pipe End Connector on the pump*

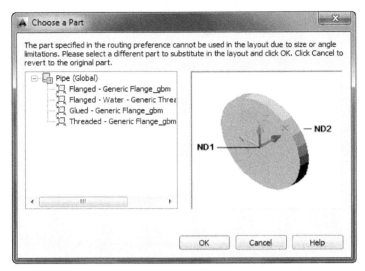

*Figure 5-51  The Choose a Part dialog box*

3.  Choose the **OK** button from the dialog box; a pipe with the length of **1000** is created. Select the Pipe End Connector, as shown in Figure 5-52; you are prompted to specify whether you want to accept the connection or you want to undo the connection made. Choose the **Accept** button from the Command prompt; the pipeline is created between Pump and Boiler. Press ENTER to exit the command.

4.  Select **Turbine** and click on the **+** sign; a rubber band pipe is attached to the cursor. Move the cursor in the left direction and enter **4000** at the Command prompt; the **Choose a Part** dialog box is displayed. Choose **OK** from the dialog box; the pipe line is created, refer to Figure 5-53.

*Figure 5-52  The pipe connector end to be selected*

*Figure 5-53* *The pipeline created*

5. Click on the Pipe End Connector, as shown in Figure 5-54; the options related to possible layouts are displayed at the Command prompt. Choose the **Next** button from the command prompt till you get the desired layout, refer to Figure 5-55.

6. Choose the **Accept** option from the command prompt; the pipe line is created between the boiler and the turbine.

*Figure 5-54* *The Pipe End Connector in the model*

7. Select the horizontal connector of the pump, move the cursor in the horizontal direction, and specify the distance as **1500** at the Command prompt; the **Choose a Part** dialog box is displayed. Choose the **OK** button from the dialog box and click on the Pipe End Connector available below the condenser; the pipe line is created, as shown in Figure 5-55.

8. Choose the **Accept** button from the Command prompt to accept the pipe line; the **Custom Sizes** dialog box is displayed. Choose the **Yes** options from the dialog box. Press ENTER to exit the tool.

9. Select the Condenser and click on the **+** sign displayed below the condenser, refer to Figure 5-56; a rubber band pipe is attached to the cursor. Now, click on the Pipe End Connector displayed on the turbine. Accept the default settings and custom parts. The pipe line is created and the final model is displayed, as shown in Figure 5-57. Press ENTER to exit the tool.

*Figure 5-55*  *The drawing after creating pipeline between condenser and pump*

*Figure 5-56*  *The Pipe End Connector to be selected*

*Figure 5-57*  *The final model*

### Saving the drawing file

1.  Choose **Save** from the **Application** menu to save the drawing file.

## Self-Evaluation Test

Answer the following questions and then compare them to those given at the end of this chapter:

1.  You need to choose the **Plumbing** option from the **Workspace Switching** flyout to activate the **Piping** workspace. (T/F)

2.  The elevation value of an equipment can be specified only from the ground. (T/F)

3.  The **Equipment** drop-down is available in the _____ panel of the **Home** tab.

4.  The cut length of a pipe can be specified in the _____ edit box.

5.  Which of the following tools is used to create a custom part ?

    (a) **Equipment**                    (b) **Content Builder**
    (c) **Catalog Editor**               (d) **Style Manager**

## Review Questions

Answer the following questions:

1.  The **Pipe Custom Fitting** tool is available in the **Pipe** drop-down of the **Build** panel. (T/F).

2.  The **Branch Angle** option is used to specify the lateral angle of a Tee at the bend. (T/F)

3.  The _____ tool is used to add valve.

4.  To create a custom Multi-view Part, you must have a _____ cf the model.

5.  Which of the following panel contains the **Cone** tool?

    (a) **Build**                        (b) **Draw**
    (c) **Block**                        (d) **Modeling**

# Exercise

## Exercise 1

In this exercise, you will create model of a piping system, as shown in Figure 5-58. Figure 5-59 shows the plan of the piping system.                              **(Expected time: 30 min)**

**Figure 5-58** *Model of the piping system*

**Figure 5-59** *Plan of the piping system*

**Answers to Self-Evaluation Test**
1. F, 2. F, 3. Build, 4. Cut length, 5. Content Builder

# Chapter 6

## Creating Plumbing System

### Learning Objectives

*After completing this chapter, you will be able to:*

- *Use equipment required in plumbing system*
- *Change basic settings of a plumbing system*
- *Configure plumbing options*
- *Route the plumbing line*
- *Create a plumbing system*

# INTRODUCTION

In this chapter, you will learn the usage of various mechanical equipment required for creating a plumbing system. A plumbing system is used to drain the waterborne waste from different locations. It consists of plumbing equipment and the plumbing line. Plumbing equipment and plumbing line together control the flow of water at various locations in a building. To create a plumbing system, first you need to figure out variables such as the vertical heads, pressure required at various locations, and the flow rate of water. These variables are used to determine the capacity of equipment to be added to the plumbing system.

# PLUMBING WORKSPACE

To create a plumbing system, you need to invoke the Plumbing Workspace first. To invoke this workspace, choose the **Workspace Switching** button from the **Application Status Bar**; a flyout will be displayed. Choose the **Plumbing** option from the flyout; the **Plumbing** workspace will be activated. The equipment that can be added while working in the **Plumbing** workspace are available in the **Equipment** drop-down of the **Build** panel in the **Home** tab of the **Ribbon**, refer to Figure 6-1. The tools available in this drop-down are discussed next.

*Figure 6-1  The Equipment drop-down*

## Filter

A filter is an equipment or a device which is used to filter the fluid before making it flow through an outlet. To add a filter in the drawing, choose the **Filter** tool from the **Equipment** drop-down available in the **Build** panel of the **Home** tab in the **Ribbon**; the **Add Multi-view Parts** dialog box will be displayed, as shown in Figure 6-2. Select the **Panel Filters** sub-category in the **Filter** category of Mechanical equipment in this dialog box; preview of the selected component is displayed on the right in the dialog box. Various sizes for the filter are available in the **Part Size Name** drop-down list of this dialog box. Select the required size from this drop-down list. You can specify the elevation value of the filter from the ground or UCS by using the **Elevation** edit box available below the **Part Size Name** drop-down list. Click in the drawing area to place the filter; a compass will be displayed below the component. Using this compass, you can rotate the component at the required angle. After specifying the angle, choose the **Close** button from the **Add Multi-view Parts** dialog box to exit.

## Pump

A pump is a mechanical device that is used to supply fluid to a desired point by using mechanical force. The method to place a pump has already been discussed in Chapter 5.

*Figure 6-2  The **Add Multi-view Parts** dialog box*

## Shower

A shower is used to convert a stream of water into spray. To add a shower in the drawing, choose the **Shower** tool from the **Equipment** drop-down available in the **Build** panel of the **Home** tab in the **Ribbon**; the **Add Multi-view Parts** dialog box will be displayed, as shown in Figure 6-3. By default, the **Rectangular Shower Stall** part is selected in the **Showers** category of the **Plumbing** equipment. Select the required part from the dialog box; various sizes available for that part are displayed in the **Part Size Name** drop-down list.

Select the required size from the drop-down list and click in the drawing area to place the shower stall; a compass will be displayed below the component. Using this compass, you can rotate the component at the required angle. After specifying the angle, choose the **Close** button from the **Add Multi-view Parts** dialog box to exit the dialog box. Figure 6-4 shows the annotated shower stall. You can apply various parameters for the fixture attached to the Muti-view Parts. The options to change these parameters are available in the **Fixture Units** tab of the **Add Multi-view Parts** dialog box, refer to Figure 6-5. These options are discussed next.

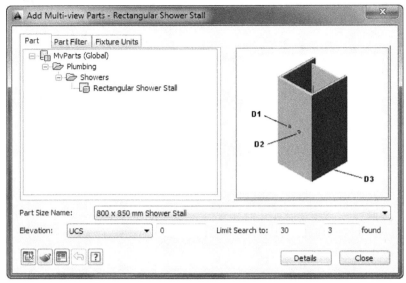

*Figure 6-3* The **Add Multi-view Parts** *dialog box*

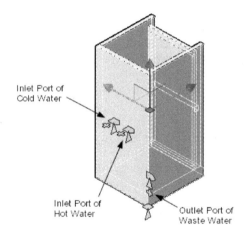

*Figure 6-4* The shower stall

## Fixture Unit Table

The **Fixture Unit Table** drop-down list is available in the **Fixture Units** tab of the **Add Multi-view Parts** dialog box. The options in this drop-down list are used to change the fixture unit table style which in turn changes options available for fixture type and occupancy type.

## Fixture

The options in this drop-down list are used to specify the type of fixture to be applied to the multi-view part.

*Figure 6-5* *The* **Add Multi-view Parts** *dialog box with the* **Fixture Units** *tab selected*

## Occupancy

The options in this drop-down list are used to specify the occupancy type of the Multi-view part. There are two options available in this drop-down list: **Public** and **Private**.

## Fixture Units for Connectors

There are various options available in this area to change properties of various connectors available for the component. The options in this area are displayed based on the part selected.

## Sink

A sink is a plumbing equipment that is used to wash utensils or any small object. There are various types of sinks available in AutoCAD MEP. These sinks are available in four main categories; **Oval Basin**, **Rectangular Basin**, **Vanity Basin**, and **Wall Mounted Basin**. To add a sink in the drawing, choose the **Sink** tool from the **Equipment** drop-down available in the **Build** panel of the **Home** tab in the **Ribbon**; the **Add Multi-view Parts** dialog box will be displayed, as shown in Figure 6-6. By default, the **Oval Basin** part is selected in the **Basins** category of the **Plumbing** equipment. Select the required part from the dialog box; various sizes available for the category are displayed in the **Part Size Name** drop-down list. Select the required size from the drop-down list and click in the drawing area to place the sink; a compass will be displayed below the component. Using this compass, you can rotate the component at the required angle. After specifying the angle, choose the **Close** button from the **Add Multi-view Parts** dialog box to exit.

*Figure 6-6* The **Add Multi-view Parts** *dialog box*

## Water Closet and Urinal

You can add water closet and urinal to the drawing area in the same way as the other equipment discussed earlier in this chapter. After choosing the **Water Closet** and **Urinal** tools from the **Equipment** drop-down, the corresponding **Add Multi-view Parts** dialog boxes will be displayed, as shown in Figures 6-7 and 6-8.

*Figure 6-7* The **Add Multi-view Parts** *dialog box displayed after choosing the* **Water Closet** *tool*

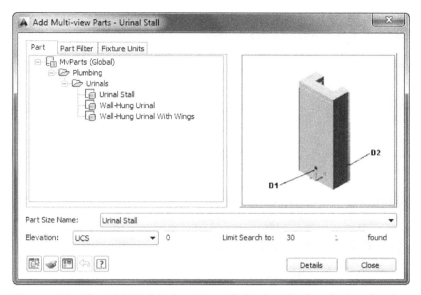

*Figure 6-8  The **Add Multi-view Parts** dialog box displayed after choosing
the **Urinal** tool*

## Equipment

When you choose the **Equipment** tool from the **Equipment** drop-down of the **Build** panel of the **Home** tab; the **Add Multi-view Parts** dialog box will be displayed, as shown in Figure 6-9. Click on the **+** sign adjacent to the desired category in the **Add Multi-view Parts** dialog box; the related equipment available in AutoCAD MEP will be displayed in a tree structure.

*Figure 6-9  The **Add Multi-view Parts** dialog box*

Select the required option from the list and click in the drawing area to place the component. A compass will also be displayed below the component. Using this compass, you can rotate the component at the required angle. After specifying the angle, choose the **Close** button from the **Add Multi-view Parts** dialog box to exit.

# PLUMBING LINE

After adding all the required equipment to the drawing, you need to add the plumbing line. To add a plumbing line, choose the **Plumbing Line** tool from the **Build** panel in the **Home** tab; you will be prompted to specify the start point of the plumbing line. Click in the drawing area to specify the start point; you will be prompted to specify the next point. Click in the drawing area to specify the next point or enter the distance at the command prompt; a plumbing line will be created and you will be prompted to specify the next point. Click in the drawing area to specify the next point or press ENTER to exit the command. The options related to plumbing line are displayed in the **Properties** palette, refer to Figure 6-10.

*Figure 6-10 The Properties palette displayed on selecting the Plumbing Line tool*

## Properties Palette

The options in the **Properties** palette are used to change the parameters of plumbing line. The options displayed in the **Properties** palette after choosing the **Plumbing Line** tool are discussed next.

### Description

This field is available in the **General** rollout of the **BASIC** rollout of the **Properties** palette in the **Design** tab. When you click on this field, the **Description** dialog box will be displayed, as shown in Figure 6-11. Enter description for the plumbing line in the **Edit the description for this object** text box in this dialog box and choose **OK** to exit.

*Figure 6-11  The Description dialog box*

## Style
This drop-down list is available in the **General** rollout of the **BASIC** rollout of the **Properties** palette. The options in this drop-down list are used to define the appearance and purpose of the plumbing line. By default, the **Standard** option is selected in the drop-down list. You can select any of the options such as **Black Pipe** and **Copper Tube - Table W**.

## System
This drop-down list is available in the **System** rollout of the **BASIC** rollout of the **Properties** palette in the **Design** tab. It contains various system definitions for the plumbing line layout.

## Nominal size
This drop-down list is available in the **Dimensions** rollout of the **BASIC** rollout. It contains options to change the size of pipe. The options available in this drop-down list change according to the option selected in the **Style** drop-down list.

## Offset
This edit box is available in the **Dimensions** rollout. Using this edit box, you can create the plumbing line by specified offset distance.

## Elevation
This edit box is available in the **Placement** rollout of the **BASIC** rollout. Using this edit box, you can specify the value of elevation of plumbing line from the ground.

## Rise
This edit box is available in the **Slope** rollout of the **BASIC** rollout in the **Properties** palette. You can specify the value of total rise in this edit box.

## Run
This edit box is available in the **Slope** rollout of the **BASIC** rollout in the **Properties** palette. You can specify the value of total run in this edit box.

## Slope (%)

This edit box is also available in the **Slope** rollout of the **BASIC** rollout in the **Properties** palette. The value in this edit box is calculated automatically depending upon the values specified in the **Run** edit box and the **Rise** edit box. If you enter the desired value in the **Slope (%)** edit box, the values in the **Rise** and **Run** edit boxes will change accordingly.

## Default Fitting Overrides

The options in this rollout are used to override the type of fitting to be chosen while creating the plumbing line. The options available in this rollout are Tee, Tee up, Tee down, and so on. To override any of the available plumbing fittings, click in the field next to the desired plumbing fitting; the **Select a style** dialog box will be displayed, refer to Figure 6-12. Select a drawing file from the **Drawing file** drop-down list; the related categories will be displayed in the **Category** drop-down list. On selecting the desired category from the drop-down list, the relevant components will be displayed in the area below the **Category** drop-down list. Select the desired component and choose the **OK** button from the dialog box; the selected component will be used as an override for the desired fitting.

*Figure 6-12  The Select a style dialog box*

## Assigned engineering ID

This drop-down list is available in the **Engineering Data** rollout of the **ADVANCED** rollout of the **Properties** palette. If the component to be added in the drawing area has an ASHRAE number, then the engineering ID for the component will be displayed in the drop-down list. You can assign any of the available engineering IDs to the component by selecting it from the drop-down list.

## Style

This drop-down list is available in the **Labels** rollout of the **Labels and Flow Arrows** rollout in the **ADVANCED** rollout of the **Properties** palette. The options in this drop-down list are used to apply the label styles to plumbing line.

The **Style** drop-down list is also available for Flow Arrows. These options have already been discussed.

# PLUMBING FITTING

The **Plumbing Fitting** tool is used to add a fitting to the plumbing line. To add a plumbing fitting, choose this tool from the **Build** panel of the **Home** tab in the **Ribbon**; you will be prompted to specify the insertion point. Also, the **Properties** palette will be displayed, as shown in Figure 6-13. Click on the plumbing line to add the plumbing fitting in the line; you will be prompted to specify the rotation value. Specify the desired rotation value at the command prompt or click in the drawing area to specify the rotation value; the plumbing fitting will be added. The **Properties** palette displayed on choosing the **Plumbing Fitting** tool is discussed next.

*Figure 6-13  The **Properties** palette displayed on choosing the **Plumbing Fitting** tool*

## Properties Palette

The options available in the **Properties** palette after choosing the **Plumbing Fitting** tool are used to change the parameters of plumbing fitting. These options are discussed next.

### Description

This field is available in the **General** rollout of the **BASIC** rollout in the **Properties** palette. When you click on this field, the **Description** dialog box will be displayed. You can write description about the fitting in the text box available in this dialog box.

### Style

This option is available in the **General** rollout of the **BASIC** rollout of the **Properties** palette. Using this option, you can specify the type of fitting to be applied to a plumbing line. To change the style, click on the field next to the **Style** option; the **Select a Style** dialog box will be displayed, as shown in Figure 6-14. Select the desired fitting from the fittings displayed in the dialog box and choose the **OK** button; the selected fitting will get attached to the cursor.

*Figure 6-14*   *The **Select a style** dialog box*

## System
This drop-down list is available in the **System** rollout of the **BASIC** rollout of the **Properties** palette. There are various options available in this drop-down list to change the application area of the fitting to be created. By default, the **Standard (STD)** option is selected in this drop-down list.

## Elevation
This edit box is available in the **Placement** rollout of the **BASIC** rollout in the **Properties** palette. Using this edit box, you can specify the value of height of the fitting from the ground level.

## Justification
This drop-down list is also available in the **Location** rollout and is used to specify the justification for plumbing line. The options in this drop-down list depend on the number of connectors available on the selected fitting. If the fitting has three connectors, then the total number of options available in this drop-down list will be four: **Insertion point**, **Connector1**, **Connector2**, and **Connector3**.

## Specify rotation on screen
This drop-down list is available in the **Location** rollout. There are two options available in this drop-down list. If you select the **Yes** option then you can specify the rotation value while adding the fitting in the drawing area. If you select the **No** option, then you need to specify the value of rotation in the edit box available in the **Properties** palette.

## Rotation

This edit box is available in the **Location** rollout only when the **No** option is selected in the **Specify rotation on screen** drop-down list. Using this edit box, you can specify the value of rotation of the fitting.

## ID

This edit box is available in the **ADVANCED** rollout of the **Properties** palette. You can specify a design ID to the fitting using this edit box.

# TUTORIAL

## Tutorial 1

In this tutorial, you will create a plumbing system, as shown in Figure 6-15. You can download the architectural file from *www.cadcim.com*. Path of the file is: *Textbooks > CAD/CAM > AutoCAD MEP > AutoCAD MEP 2014 for Designers > Tutorial.*                    **(Expected time: 30 min)**

***Figure 6-15*** *The plumbing system to be created*

The following steps are required to complete this tutorial:

Examine the model to determine the equipment to be added and the parameters of plumbing line.

a.   Open the drawing downloaded from the website.
b.   Add the equipment according to the model.
c.   Create plumbing line between various equipment.

## Downloading and Opening the Drawing File

1. Download the *c06_amep_prt.zip* file from *http://www.cadcim.com*. The path of the file is as follows: *Textbooks > CAD/CAM > AutoCAD MEP > AutoCAD MEP 2014 for Designers > Input*

2. Extract this file to desired location.

3. Open the drawing file *c06_amep_prt.dwg* from the specified location by double-clicking on it. The drawing file is displayed, as shown in Figure 6-16.

***Figure 6-16*** *The downloaded drawing file*

## Adding the Equipment

1. Change the workspace to **Plumbing** and choose the **Shower** tool from the **Equipment** drop-down available in the **Build** panel of the **Home** tab in the **Ribbon**; the **Add Multi-view Parts** dialog box is displayed, as shown in Figure 6-17. Also, the shower stall is displayed, attached to the cursor.

2. Specify **800 x 850 mm Shower Stall** in the **Part Size Name** drop-down list and **0** in the **Elevation** edit box in the dialog box.

3. Place the shower stall in the bathroom, aligned to the corner, refer to Figure 6-18.

4. Choose the **Sink** tool from the **Equipment** drop-down; the **Add Multi-view Parts** dialog box is displayed, as shown in Figure 6-19.

*Figure 6-17* The **Add Multi-view Parts** *dialog box*

*Figure 6-18* The drawing after adding equipment

5. Select **Vanity Basin** from the left area of the dialog box, and then select part size as 750mmx750mm. Click in the drawing area to place the basin, refer to Figure 6-20. Make sure the elevation value is 1000.

*Figure 6-19* The **Add Multi-view Parts** *dialog box displayed after choosing the* **Sink** *tool*

*Figure 6-20* The drawing after adding equipment

6.  Similarly, add **Bidet** available in the **Add Multi-view Parts** dialog box displayed on choosing the **Water Closet** tool from the **Equipment** drop-down, refer to Figure 6-20. Make sure the elevation value is 0.

7.  Choose the **Sink** tool from the **Equipment** drop-down and then choose **Wall Mounted Basin** from the left area; the **Add Multi-view Parts** dialog box is displayed, as shown in Figure 6-21.

**Figure 6-21** *The* **Add Multi-view Parts** *dialog box displayed after selecting* **Wall Mounted Basin**

8. Place the basins in the drawing area at elevation 1000, refer to Figure 6-22.

**Figure 6-22** *The drawing area after adding the basins*

## Creating Plumbing Line Between Various Equipment

There are three plumbing lines to be added in the system: Waste (WP), Domestic Cold Water, and Domestic Hot Water.

1. Choose the **Waste (WP)** tool from the **Tool Palettes - Plumbing** displayed at the right in AutoCAD MEP window, refer to Figure 6-23; you are prompted to select the starting point for Waste plumbing line.

2. Select the Waste Pipe End Connector of the shower stall; the other end of waste plumbing line gets attached to the cursor and you are prompted to specify the next end point.

3.   Select the Pipe End Connector of the vanity basin placed in the Kitchen area; the **Select Connector** dialog box is displayed, as shown in Figure 6-24.

4.   Choose the **Connector 3: Waste** option from the dialog box and choose the **OK** button; the **Plumbing Line - Elevation Mismatch** dialog box is displayed, as shown in Figure 6-25.

*Figure 6-23*  *The **Tool Palettes - Plumbing***

*Figure 6-24*  *The **Select Connector** dialog box*

*Figure 6-25*  *The **Plumbing Line - Elevation Mismatch** dialog box*

5. Choose the **Adjust the slope** option from the dialog box; the slope of the plumbing line is automatically adjusted to permit the flow of water and a plumbing line is created.

6. Similarly, create a plumbing line from the waste line of vanity basin to waste line of bidet in the Bathroom area and then to the waste lines of the wall mounted basins. The drawing after adding all the waste plumbing lines is displayed, as shown in Figure 6-26.

*Figure 6-26  The drawing after adding all the waste plumbing lines*

7. Choose the **Domestic Hot Water** tool from the **Tool Palettes - Plumbing** and add the hot water plumbing line, as shown in Figure 6-27 (displayed in red color).

*Figure 6-27  The drawing after adding all the hot water plumbing lines*

8.  Choose the **Domestic Cold Water** tool from the **Tool Palettes - Plumbing** and add the cold water plumbing line, as shown in Figure 6-28 (displayed in blue color).

*Figure 6-28  The drawing after adding all the cold water plumbing lines*

The drawing after adding all the plumbing lines is displayed, as shown in Figure 6-29.

## Saving the drawing file

1.  Choose **Save** from the **Application** menu to save the drawing file.

*Figure 6-29  The drawing after adding all the plumbing lines*

## Self-Evaluation Test

**Answer the following questions and then compare them to those given at the end of this chapter:**

1. You cannot change the type of fitting after adding it in the plumbing line. (T/F)

2. Rotation of a fitting can be specified either by using an edit box or by dynamically rotating it. (T/F)

3. The value in the **Slope** edit box is calculated automatically depending upon the values specified in the _____ and _____ edit boxes.

4. The _____ drop-down list in the **Properties** palette is used to specify the type of system for plumbing line.

5. Which of the following equipment does not require a fixture?

   (a) **Filter**            (b) **Shower**
   (c) **Sink**              (d) **Pump**

## Review Questions

**Answer the following questions:**

1. The **Pipe Custom Fitting** tool is not available in the **Plumbing** workspace. (T/F)

2. The **End Cap** option is available in the **Select a style** dialog box. (T/F)

3. The **Domestic Hot Water** tool is available in the **Tool Palettes -** _____.

4. To create a custom multi-view part, you must have a _____ of the model in the current drawing.

5. Which of the following is not a type of plumbing fitting?

   (a) **Tee**              (b) **Cross**
   (c) **Elbow**            (d) **Wye**

# Exercise

## Exercise 1

In this exercise, you will create the model of a plumbing system using the drawing shown in Figure 6-30. You can download the architectural file from *www.cadcim.com*. Path of the file is: *Textbooks > CAD/CAM > AutoCAD MEP > AutoCAD MEP 2014 for Designers > Tutorial.*

**(Expected time: 30 min)**

*Figure 6-30  The drawing of plumbing system*

# Chapter 7

# Creating Electrical System Layout

## Learning Objectives

*After completing this chapter, you will be able to:*

- *Use equipment required in Electrical System*
- *Change basic settings of an Electrical System*
- *Configure electrical options*
- *Create wires*
- *Create a cable tray*
- *Create an electrical panel*
- *Create a cable tray fitting*
- *Add conduit fittings*
- *Calculate total load of devices*
- *Calculate loads and wire size*

# INTRODUCTION

In this chapter, you will learn the usage of various electrical equipment for creating an electrical system. An electrical system is used to transmit power from one location to the other. It is composed of equipment, panels, cable trays, wires, conduits and so on. For creating an electrical system, choose the **Electrical** option from the **Workspace Switching** flyout, the **Electrical** workspace will be activated. The equipment available in the **Electrical** workspace of AutoCAD MEP are discussed next.

# ADDING EQUIPMENT

For creating an electrical system, you need to add related equipment to the structure. All the equipment that can be added while working in the **Electrical** workspace, are available in the **Equipment** drop-down of the **Build** panel in the **Home** tab of the **Ribbon**, refer to Figure 7-1. Some of the options available in this drop-down are discussed next.

*Figure 7-1  The Equipment drop-down*

## Generator

A generator is an equipment or a device, which is used to convert mechanical energy into electrical energy. To add a generator to the drawing, choose the **Generator** tool from the **Equipment** drop-down of the **Build** panel in the **Home** tab of the **Ribbon**; the **Add Multi-view Parts** dialog box will be displayed, as shown in Figure 7-2. By default, **200-600kW Emergency Power Generator - Diesel** is selected in the **Part** tab. Preview of the selected component is displayed in the right of the dialog box. Various sizes for the generator, depending on their load capacity, are available in the **Part Size Name** drop-down list of this dialog box. Select the required size from this drop-down list. Click in the drawing area to place the generator; a compass will be displayed below the component. Using this compass, you can rotate the component at a specific angle. After specifying the angle, choose the **Close** button from the **Add Multi-view Parts** dialog box to exit the dialog box.

You can also specify the elevation value of the Generator from the ground or UCS by using the **Elevation** edit box available below the **Part Size Name** drop-down list.

## Junction Box

A junction box is a metallic or plastic container that is used to hide the electrical connections. To add a junction box, choose the **Junction Box** tool from the **Equipment** drop-down of the **Build** panel in the **Home** tab of the **Ribbon**; the **Add Multi-view Parts** dialog box will be displayed, as shown in Figure 7-3. Select the desired junction box from the **Junction Boxes** node in the **Part** tab of the dialog box. Various sizes of the selected junction box are displayed in the **Part Size Name** drop-down list. By default, 13 Hole Large Outlet Boxes is selected in the **Part** tab. Select the required size from the drop-down list. Now, click on the drawing area to place the junction box; a compass will be displayed below the junction box. Using this compass, you can rotate the component at a specific angle. After specifying the angle, choose the **Close** button from the **Add Multi-view Parts** dialog box to exit the dialog box.

*Figure 7-2* The **Add Multi-view Parts** *dialog box with the* **200-600 kW Emergency Power Generator - Diesel** *part selected*

*Figure 7-3* The **Add Multi-view Parts** *dialog box with the* **13 Hole Large Outlet Boxes** *part selected*

## Switchboard

A switchboard is an arrangement of electrical switches connected to each other in a close packed unit. A switch board controls the power of various areas of a serving unit. To add a switchboard, choose the **Switchboard** tool from the **Equipment** drop-down of the **Build** panel in the **Home** tab of the **Ribbon**; the **Add Multi-view Parts** dialog box will be displayed, as shown in Figure 7-4. Expand the **Switchboard** node and click on the desired switchboard type; the preview of the selected switch board will be displayed in the preview area of the dialog box. You can select the desired size from the **Part Size Name** drop-down list displayed below the preview area of the dialog box. Now, click in the drawing area to place the switchboard; a compass will be displayed below the component. Using this compass, you can rotate the component at a specific angle. After specifying the angle, choose the **Close** button from the **Add Multi-view Parts** dialog box to exit the dialog box.

*Figure 7-4* *The* ***Add Multi-view Parts*** *dialog box with the* ***Circuit Breaker Switchboard*** *part selected*

There are three types of switchboards available in AutoCAD MEP by default: circuit breaker switchboard, distribution board, and utility switchboard. The circuit breaker switchboard consists of switches that are automatically operated to protect the circuit from the damage caused by overload or short circuit. The distribution boards are used to distribute the power coming from source to various outlets. The utility switchboards are used to transmit power for a specific purpose.

The other tools in this drop-down also work in the similar way as discussed above.

# PANEL

Panels are the containers to hold various switches arranged for a specific purpose. To add a panel to the drawing, choose the **Panel** tool from the **Build** panel of the **Home** tab in the **Ribbon**; the preview of the panel attached to the cursor will be displayed, refer to Figure 7-5. The **Properties** palette on choosing this tool will be displayed, as shown in Figure 7-6. Now, click in the drawing area to place the panel. You will be prompted to specify the rotation value for the panel. Specify the rotation value and press ENTER; the panel will be created at the specified location.

*Figure 7-5* *Preview of the panel*

*Figure 7-6* *The **Properties** palette displayed on choosing the **Panel** tool*

The options available in the **Properties** palette after choosing the **Panel** tool are discussed next.

**Note**
*The symbols in the **Style** field and the **Select a style** dialog box are displayed with black background if you have set the background color of drawing area as black.*

## Description

This option is used to specify the description about the object. When you click in the field corresponding to this option, the **Description** dialog box will be displayed, as shown in Figure 7-7. You can enter the description about the component in the **Edit the description for this component** text box of the dialog box. After specifying the description, choose the **OK** button to exit the dialog box.

*Figure 7-7* The **Description** *dialog box*

## Style

This option is used to change the style of the panel to be created. Click in the field corresponding to this option; the **Select a style** dialog box will be displayed, as shown in Figure 7-8. Select the desired category from the **Category** drop-down list in the dialog box; the styles available in the selected category will be displayed in the dialog box. Select the desired style from the dialog box and choose the **OK** button; the selected style will be assigned to the panel.

*Figure 7-8* The **Select a style** *dialog box*

## Align to objects
The options in this drop-down list are used to specify whether the panel will be aligned to the selected object or not. There are two options available in this drop-down list: **Yes** and **No**. The **Yes** option is used to specify that the panel will be aligned to the selected object.

## Rotation
This option is used to specify the rotation angle value of the panel.

## Justification
This drop-down list is available in the **Properties** palette only when the **No** option is selected in the **Align to Objects** drop-down list. The options in this drop-down list are used to specify the position of the panel with respect to the insertion point.

## Preset elevation
The options in this drop-down list are used to set the elevation of the panel. By default, **UCS** is selected in this drop-down list.

## Elevation
The option is used to specify the value of elevation from the selected preset.

## System
The options in this drop-down list are used to specify the type of system for which the panel is being created.

## Create circuits
The options in this drop-down list are used to specify whether to create a circuit for the panel or not. There are two options available in this drop-down list: **Yes** and **No**. If you choose the Yes option from this drop-down list, then the circuit will also be created with the panel.

## Circuit Settings
On clicking in this field, the **Circuit Settings** dialog box will be displayed, as shown in Figure 7-9. Using the options in this dialog box, you can define various settings for the circuit. Choose the **OK** button after defining the settings.

## Name
The **Name** edit box is used to specify the name of the panel.

## Rating
This edit box is used to specify the rating of the panel.

## Voltage phase-to-neutral
The options in this drop-down list are used to specify the voltage value between phase line and neutral line.

**Figure 7-9** *The **Circuit Settings** dialog box*

# Voltage phase-to-phase

The options in this drop-down list are used to specify the voltage value between two phase lines.

## Phases

The options in this drop-down list are used to specify whether the selected panel is for single phase supply or three phase supply. There are two options available in this drop-down list: **1** and **3**. If you select the **1** option from this drop-down list, the panel will be created for single phase supply. If you select the **3** option from this drop-down list, the panel will be created for three phase supply.

## Wires

The options in this drop-down list are used to specify the number of cables to be attached with the panel after creation. If the **1** option is selected from the **Phases** drop-down list, only one option will be available in this drop-down list. If the **3** option is selected in the **Phases** drop-down list, then the **3** and **4** options will be available in this drop-down list.

**Note**
*If the 3 option is chosen in both the **Phases** and the **Wires** drop-down lists, then the **Voltage phase-to-neutral** edit box will not be available in the **Properties** palette.*

## Main type

There are two options available in this drop-down list: the **Main lugs only (MLO)** and the **Main circuit breaker (MCB)**. These options are used to specify whether the panel is of main lug only or main circuit breaker.

## Main size (amps)

This edit box is used to specify the value of current running through the main supply.

## Design capacity (amps)

This edit box is used to specify the designed capacity of the panel.

## Panel type

The options in this drop-down list are used to specify the type of panel to be used. There are two options available in this drop-down list: **ANSI** and **ISO**.

## Enclosure type

This edit box is used to specify the type of enclosure for the panel.

## Mounting

The options in this drop-down list are used to specify the type of mounting required for the panel. There are three options available in this drop-down list: **Surface**, **Recessed**, and **Floor**.

## AIC rating

The option is used to specify the short circuit rating. You can specify the maximum value of current that can be borne in the situation of a short circuit.

## Fed from

This option is used to specify the source of power for the current panel.

## Notes

This option is used to specify important notes that are to be taken care of while handling the panel.

## DEVICE

This tool is used to insert an electrical device in the drawing area. This tool is available in the **Build** panel of the **Home** tab in the **Ribbon**. To insert an electrical device, choose the **Device** tool; you will be prompted to specify an insertion point for the device. Also, the **Properties** palette will be displayed, as shown in Figure 7-10. Click in the drawing area;

the device will be placed at the specified point. The options available in the **Properties** palette after choosing the **Device** tool are discussed next.

*Figure 7-10  The **Properties** palette displayed on choosing the **Device** tool*

## Description
This option is used to enter the description of the device to be added.

## Style
This option is used to select the type of device to be inserted in the drawing area. To select the type of device, click in the field available for this option; the **Select a Style** dialog box will be displayed, as shown in Figure 7-11.

Choose the desired options from the **Drawing file** and **Category** drop-down lists; the relevant device symbols will be displayed in the dialog box. Select the desired device and choose the **OK** button. The selected device will get attached to the cursor and will be displayed in the **Style** field of the **Properties** palette.

## Layout method
The options in this drop-down list are used to specify the method of insertion of the device. There are three options available in this drop-down list: **One by one**, **Distance around space**, and **Quantity around space**. If you select the **One by one** option from this drop-down list then you need to insert the devices one by one. If the **Distance around space** option is selected then

*Figure 7-11  The Select a style dialog box*

you can specify the distance around the boundary by using the edit boxes displayed below it. On selecting the **Distance around space** option, the **Distance between** and the **Number of devices** edit boxes will be displayed below the option. The **Distance between** and the **Number of devices** edit boxes cannot be activated at the same time. On selecting the **Distance around space** option, only the **Distance between** edit box is activated. If you choose the **Quantity around space** option from the drop-down list; the **Distance between** edit box will be deactivated whereas the **Number of devices** edit box will be activated. You can specify the number of devices in the **Number of devices** edit box. Note that to use both the **Distance around space** and the **Quantity around space** options, you must have a space created in the model.

## Align to objects

The options in this drop-down list are used to specify whether the device is to be aligned to the selected object or not. There are two options available in this drop-down list: **Yes** and **No**. Choose the **Yes** option from this drop-down list to align the device with the selected object.

## Rotation

This option is used to specify the rotation angle value for the device.

## Justification

The options in this drop-down list are used to justify the device with respect to the insertion point.

## Preset Elevation

The options in this drop-down list are used to set elevation of the device. By default, **UCS** is selected in this drop-down list.

## Elevation

This option is used to specify the value of elevation from the selected preset.

## System

The options in this drop-down list are used to specify the system type in which the device is inserted.

## ID

This option is used to assign an ID to the current device.

## Insert tag

The options in this drop-down list are used to specify the tag to be added to the selected device.

## Electrical Property

On clicking in this field, the **Electrical Properties** dialog box will be displayed, as shown in Figure 7-12. While adding a new device, you can change the number of connectors and other related properties by using the options available in this dialog box.

## CABLE TRAY

This tool is used to add a cable tray for supporting the cables. To add a cable tray, choose the **Cable Tray** tool from the **Build** panel in the **Home** tab of the **Ribbon**; you will be prompted to specify start point of the cable tray. Also, the **Add Cable Trays** dialog box will be displayed, as shown in Figure 7-13. Now, click in the drawing area to specify the start point of the cable tray; you will be prompted to specify the end point of the cable tray. Click to specify the end point of the cable tray; you will be prompted to specify the end point of the cable tray again. You can specify the end point of the next section or you can press ENTER to exit. The options available in this dialog box are discussed next.

## System

The options in this drop-down list are used to specify the system type in which the cable tray is being added.

## Elevation

There are two options available for specifying elevation in the dialog box; the **Elevation** drop-down and **Elevation** edit box. The **Elevation** edit box is used to specify the value of elevation of the cable tray from the reference. The options in the **Elevation** drop-down list are used to specify reference for the elevation.

*Figure 7-12* The **Electrical Properties** *dialog box*

*Figure 7-13* The **Add Cable Trays** *dialog box*

# Horizontal

There are two options available to horizontally position the cable tray with respect to the insertion point: the **Justification** drop-down list and the **Offset** edit box. The options in the **Justification** drop-down list are used to specify the justification method for the cable tray in horizontal direction. The **Offset** edit box is used to specify the distance of the cable tray from the cursor.

## Vertical

There are two options available to vertically position the cable tray with respect to the insertion point: the **Justification** drop-down list and the **Offset** edit box. The options in the **Justification** drop-down list are used to specify the justification method for the cable tray in vertical direction. The **Offset** edit box is used to specify the distance of the cable tray from the cursor.

## Use Rise/Run

This radio button is used to specify the rise/run value for the cable tray. On selecting this radio button, an edit box adjacent to this radio button will be activated. You can specify the value for rise and run in this edit box.

## Use Routing

By default, this radio button is selected in the **Layout Method** area. As a result, the **Elbow Angle** edit box will be activated. You can specify the elbow angle value in this edit box.

## CABLE TRAY FITTING

This tool is used to add a user defined cable tray fitting. To create a cable tray fitting, choose the **Cable Tray Fitting** tool from the **Build** panel in the **Home** tab of the **Ribbon**; the **Add Cable Tray Fitting** dialog box will be displayed, as shown in Figure 7-14. By default, the last used cable tray fitting is selected in this dialog box. Select the desired part from the **Part** tab of the dialog box; preview of the selected cable tray fitting will be displayed in the right of the dialog box and the selected part will get attached to the cursor. Select the desired size of the selected part from the **Part Size Name** drop-down list in the dialog box and click in the drawing area to place the cable tray fitting. You can create a cable tray fitting as a separate entity or can join it with the cable tray.

## WIRE

This tool is used to add a wire in the drawing area. To do so, choose the **Wire** tool from the **Build** panel of the **Home** tab in the **Ribbon**; you will be prompted to specify the start point of the wire and the corresponding **Properties** palette will be displayed, as shown in Figure 7-15. Click in the drawing area to specify the first point of the wire; you will be prompted to specify the second point of the wire. Click to specify the second point of the wire; the wire will be created. You can specify more points or you can press ENTER to exit the tool.

**Tip**
*You can also convert a line into wire. To do so, select the line and right-click on it; a shortcut menu will be displayed. Hover the cursor over the **Convert To** option and choose the **Wire** option from flyout displayed; you will be prompted to delete the line after creating the wire or not. Enter **Y** or **N**; the wire will be converted.*

*Figure 7-14 The **Add Cable Tray Fittings** dialog box*

*Figure 7-15 The **Properties** palette displayed on choosing the **Wire** tool*

The options available in the **Properties** palette, after choosing the **Wire** tool, are discussed next.

## Description

This option is used to specify description about the wire.

## Style

The options in this drop-down list are used to specify the wire type.

## Segment

The options in this drop-down list are used to specify the segment type to be used while creating the wire system. There are six options available in this drop-down list: **Line**, **Arc**, **Snake**, **Polyline**, **Chamfer**, and **Spline**. On choosing an option from this drop-down list, the preview of the corresponding segment type is displayed.

## Height

This option is available below the **Segment** drop-down list only when the **Arc**, **Snake**, **Chamfer**, or the **Spline** option is selected in the **Segment** drop-down list. It is used to specify the height of the segment.

## Offset

This drop-down list is available below the **Height** edit box only when the **Arc**, **Snake**, **Chamfer**, or the **Spline** option is selected in the **Segment** drop-down list. The options in this drop-down list are used to specify the alignment of the segment. There are two options available in this drop-down list: **Left** and **Right**. The **Left** option is used to align the segment to the left. The **Right** option is used to align the segment to the right.

## Radius

This option is available only when the **Polyline** option is selected in the **Segment** drop-down list. This option is used to specify the radius at corners of the polyline segment.

## Preset Elevation

The options in this drop-down list are used to set the value for elevation. By default, **UCS** is selected in this drop-down list.

## Elevation

This option is used to specify the value of elevation from the selected preset.

## System

The options in this drop-down list are used to specify the system type in which the wire will be added.

## Show circuits from the panels

The options in this drop-down list are used to specify whether you want to display the circuits from the panels or not.

## Circuit

The options in this drop-down list are used to specify the circuit for which the wire is being created.

## Connected circuits

This option is used to specify the circuits that are connected to wire.

## Connected load

This option is used to specify the total load in the circuit connected to the wire.

## Hot size

This option is used to specify the size of the hot wire in the wiring system.

## Neutral size

This option is used to specify the size of the neutral wire in the wiring system.

## Ground size

This option is used to specify the size of the ground wire in the wiring system.

## New Run

This tool is used to start a new run of wire system. This tool is available at the bottom of the **Properties** palette. It will be active after making one run of wire. On choosing this tool, you can start creating a new wiring system by specifying the start point.

# CONDUIT

 The **Conduit** tool is used to add conduits to the drawing area. To do so, choose this tool from the **Conduit** drop-down of the **Build** panel in the **Home** tab of the **Ribbon**; the **Properties** palette will be displayed, as shown in Figure 7-16 and you will be prompted to specify the start point of the conduit. Click in the drawing area to specify the start point of the conduit; you will be prompted to specify the next point of the conduit. Click to specify the end point of the conduit; you will be again prompted to specify the next point of the conduit. You can specify the next point or you can press ENTER to exit the tool. Most of the options in this palette have already been discussed. Remaining options are discussed next.

## Routing preference

The options in this drop-down list are used to specify the preferred routing system for the conduit creation.

*Figure 7-16  The **Properties** palette displayed on choosing the **Conduit** tool*

## Nominal size
This option is used to specify the diameter of the conduit pipe.

## Specify cut length
The options in this drop-down list are used to specify whether the conduit has a cut length or not.

## Cut length

This option is available only when the **Yes** option is selected in the **Specify cut length** drop-down list. It is used to specify the maximum length of one segment of the conduit.

## Justify

The options available in this drop-down list are used to specify the justification method for the conduit.

## Slope format

The options in this drop-down list are used to specify the format in which the slope value will be entered. There are four options available in this drop-down list: **Angle (Decimal Degree)**, **Percentage (100%=45)**, **Percentage (100%=90)**, and **Rise value/Run value**.

## Slope

This edit box is used to specify the value of slope. The format of the slope value specified in this edit box will depend on the option selected in the **Slope format** drop-down list. If you change the slope value while adding more ducts, then the fitting at joints will adjust accordingly.

## Bend Angle

This drop-down list is used to specify the angle of bend that is to be applied on a bend while creating conduit.

## Bend Radius

This edit box is used to specify the radius of bend that is to be applied on a bend while creating conduit.

## Connection details

When you click in this field; the **Connection Details** dialog box will be displayed, as shown in Figure 7-17. There are two rollouts available in the dialog box containing the details about both the connections.

## Preferences

When you click in this field, the **Conduit Layout Preferences** dialog box will be displayed, as shown in Figure 7-18. The options in this dialog box are used to define the layout preferences for creating a conduit. Using the options in this dialog box, you can set the preferences for Slope, Elevation, Label style, Flow arrow style, and other parameters.

*Figure 7-17*  *The* **Connection Details** *dialog box*

*Figure 7-18*  *The* **Conduit Layout Preferences** *dialog box*

## Style

There are two drop-down lists with the name Style; one in the **Labels** rollout and other in the **Flow Arrows** rollout of the **Labels and Flow Arrows** rollout in the **ADVANCED** rollout of the **Properti**es palette. Options in these drop-down lists are used to apply the styles to labels and flow arrows of the conduit.

# PARALLEL CONDUITS

The **Parallel Conduits** tool is used to add parallel conduits to the drawing area. To do so, choose this tool from the **Conduit** drop-down in the **Build** panel of the **Home** tab in the **Ribbon**; you will be prompted to select baseline objects. Select the baseline objects, refer to Figure 7-19; you will be prompted to select the parallel conduits. Select the conduits parallel to the selected baseline and press ENTER; you will be prompted to specify the next point for the conduit. Click in the drawing area to specify the next point of the conduit; a parallel conduit consisting of the selected conduit lines is displayed, as shown in Figure 7-20.

*Figure 7-19 The baseline objects to be selected*

*Figure 7-20* *The parallel conduits created*

# CONDUIT FITTING

Conduit Fitting  The **Conduit Fitting** tool is used to add fittings to the conduit. To add a conduit fitting, choose this tool from the **Build** panel in the **Home** tab of the **Ribbon**; you will be prompted to specify an insertion point. Also, the **Properties** palette will be displayed, as shown in Figure 7-21. Click in the drawing area or on a conduit line to add the conduit fitting. The options in the **Properties** palette after choosing the **Conduit Fitting** tool are discussed next.

## Description

This option is used to specify description about the conduit fitting.

## System

The options in this drop-down list are used to specify the system for which the conduit is being added.

*Figure 7-21  The **Properties** palette displayed after choosing the **Conduit Fitting** tool*

## Part
This option is used to select the required fitting type to be used. Click in the field next to the **Part** option in the Properties palette; the **Select a part** dialog box will be displayed, as shown in Figure 7-22. Select the desired part from the dialog box and choose the **OK** button; the selected conduit fitting gets attached to the cursor. The options available in this dialog box are discussed next.

## Type
The options in this drop-down list are used to specify the part type to be used as conduit fitting. There are five options available in this drop-down list: **All**, **Conduit Body**, **Elbow**, **Tee**, and **Transition**. Choose any of the option from the drop-down list; the related options are displayed in the dialog box.

## Subtype
The options in this drop-down list are used to choose the subtypes available for the selected type. On choosing a subtype from the drop-down list, the related options are displayed in the **SELECT PART FROM CATALOG** area of the dialog box.

## SELECT PART FROM CATALOG

The parts available for selected type and subtype are displayed in this area. Choose the required part from this area and then choose the **OK** button from the dialog box; the selected part gets attached to the cursor.

*Figure 7-22* The **Select a part** dialog box

## FILTER

The options in this area are used to filter the parts available for selected type and subtype depending on their size and parameters.

## SELECT PART SIZE

The options in this area are used to select required size of the selected part. The options displayed in this area are confined by the values selected in the **Filter** area of the dialog box. Select the part having required size from this area and choose the **OK** button from this dialog box; the part will get attached to the cursor.

# Current Size

The options in this drop-down list are used to specify the size for the selected part. These options are also available in the **SELECT PART SIZE** area of the **Select a part** dialog box which has been discussed earlier.

## Nominal Diameter

This edit box is available for each connector added to the part. Using this edit box, you can change the nominal diameter of the connector.

## Other Dimensions Rollout

The options in this rollout are used to specify various dimensions for the part. The options in this rollout change depending upon the selected part.

## Elevation

This edit box is used to specify the value of elevation for the selected part.

## Specify rotation on screen

The options in this drop-down list are used to specify whether the rotation of the part will be specified on the screen or in the edit box available in the **Properties** palette. If you choose the **No** option from the drop-down list; the **Rotation** edit box will be displayed.

## Rotation

You can specify the value of rotation in this edit box.

## Connection details

This button is available in the **ADVANCED** rollout. When you choose this button, the **Connection Details** dialog box will be displayed, as shown in Figure 7-23. There are two rollouts available in the dialog box, containing the details about both the ends of the conduit.

*Figure 7-23* The **Connection Details** *dialog box*

# CIRCUIT MANAGER

In an electrical layout, all the appliances are connected to the power sources with the help of a circuit. In AutoCAD MEP, circuits are used for calculating circuit loads, checking circuit overloads, or calculating wire sizes. There can be more than one circuit in a single electrical layout. To create and manage circuits in AutoCAD MEP, the Circuit Manager is used. To invoke the **Circuit Manager**, choose the **Circuit Manager** tool from the **Electrical** panel in the **Analyze** tab of the **Ribbon**; the **Circuit Manager** will be displayed, as shown in Figure 7-24. The options available in the **Circuit Manager** are discussed next.

*Figure 7-24  The Circuit Manager*

## Create New Circuit

This drop-down is available at the bottom of the **Circuit Manager**. The tools in this drop-down are used to create new circuits to supply power to appliances depending on the required load, refer to Figure 7-25.  The tools available in this drop-down are discussed next.

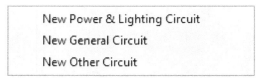

*Figure 7-25  The Create New Circuits drop-down*

### New Power & Lighting Circuit

This tool is used to create a circuit for power and lighting. To create such type of circuits, choose the **New Power & Lighting Circuit** tool from the **Create New Circuits** drop-down; the **AutoCAD MEP - Electrical Project Database** dialog box will be displayed, as shown in Figure 7-26 and you will be prompted to create a new electrical project database (EPD) or open an existing database. Choose the **Create a new EPD file** option from the dialog box; the **Save As** dialog box will be displayed, as shown in Figure 7-27. Specify the desired file name and choose the **Save** button from the dialog box. A circuit will be created and then displayed under the **Power and Lighting** node in the **Circuit Manager**. Also, the related parameters will be displayed in the right side of the dialog box. You can change the parameters as per your requirement.

*Figure 7-26  The AutoCAD MEP - Electrical Project Database dialog box*

*Figure 7-27  The Save As dialog box*

## New General Circuit

This tool is used to create general circuits. General circuits are used for general application. Figure 7-28 shows a circuit created by using the **New General Circuit** tool in the **Circuit Manager**. There are four parameters available in the right side of the **Circuit Manager** for a general circuit. You can change these parameters as per your requirement.

*Figure 7-28* *The* *Circuit Manager* *with the General circuit*

## New Other Circuit

This tool is used to create new circuits for special purposes like a circuit to run air conditioner. The working of this tool is similar to the **New General Circuit** tool.

# Create Multiple Circuits

This tool is used to create multiple circuits at a time. To do so, choose the **Create Multiple Circuits** tool available at the bottom of the **Circuit Manager**; the **Create Multiple Circuits** dialog box will be displayed, as shown in Figure 7-29. Select the desired options and choose the **OK** button from the dialog box to create multiple circuits. The options available in this dialog box are discussed next.

## Panel

The options in this drop-down list are used to select the panel to be created. All the panels created in the current drawing are displayed in this drop-down list.

## System Type

The options in this drop-down list are used to select the system type for the circuits to be created. There are three options available in this drop-down list: **Power and Lighting**, **General**, and **Other**.

## System

The options in this drop-down list are used to specify the system for which the circuit is to be created. The options in this drop-down list change according to the option selected in the **System Type** drop-down list.

## Total number of slots

This option is used to specify the number of slots required for current electrical layout. You cannot change the value of this spinner directly. The value in this spinner is changed automatically depending on the values selected in the **Number of circuits** spinner if the **Other** option or the **General** option is selected in the **System Type** drop-down list. If the **Power and**

**Lighting** option is selected in the **System Type** drop-down list, then the value in the **Total number of slots** changes according to the values specified in the **Number of 3-pole circuits**, **Number of 2-pole circuits**, and **Number of 1-pole circuits** spinners.

*Figure 7-29  The **Create Multiple Circuits** dialog box*

## Number of 3-pole circuits

This option is available only when the **Power and Lighting** option is selected in the **System Type** drop-down list, refer to Figure 7-30. Using this option, you can specify the total number of circuits that will use three pole electric supply.

## Number of 2-pole circuits

This option is also available only for Power and Lighting system type. Using this option, you can specify the total number of circuits that will use the two pole electric supply.

## Number of 1-pole circuits

This option is also available only for Power and Lighting system type. Using this option, you can specify the total number of circuits that will use the one pole electric supply.

## Voltage

These options are also activated only when the **Power and Lighting** option is selected in the **System Type** drop-down list. The **Voltage** options are available for 1-pole circuits, 2-pole circuits, and 3-pole circuits. You can change the value of voltage for a specific circuit by using this spinner, refer to Figure 7-30.

*Figure 7-30  The **Create Multiple Circuits** dialog box with circuits selected*

## Description for circuits with breakers

This edit box is used to specify the default description about the circuit having circuit breakers.

## Description for circuits without breakers

This edit box is used to specify the default description about the circuits having no circuit breaker.

# Delete Circuit

This tool is used to delete circuits that are created earlier. To delete a circuit, select the circuit to be deleted from the left area and then choose the **Delete Circuit** tool from the bottom of the **Circuit Manager**; the selected circuit will be deleted.

## Show Circuited Devices

This tool is used to display the devices that are attached to the current circuit. On choosing this tool; the devices attached to the current circuit are highlighted in the drawing area.

## Circuit Report

This tool is used to display a report for the current circuit containing the data related to its various parameters. To display a circuit report, choose this tool from the bottom of the **Circuit Manager**; the **Circuit Report** dialog box will be displayed, refer to Figure 7-31. This report consists of all the information for a circuit such as load, voltage, and length.

*Figure 7-31* The **Circuit Report** dialog box

## Cut Circuit

This tool is used to remove a circuit from a panel system and place it in another panel system. To do so, select the circuit from one panel and choose this tool from the bottom of the **Circuit Manager**, refer to Figure 7-32.

*Figure 7-32* The **Circuit Manager** with the circuit to be shifted

## Copy Circuit

This tool is used to copy the selected circuit. The circuit selected is copied to the clipboard and then it can be pasted in any of the panels. To copy a circuit, select the circuit from the left area of the **Circuit Manager** and then choose the **Copy Circuit** tool from the toolbar available at the bottom; the circuit will be copied to the clipboard.

## Paste Circuit

This tool is used to paste an already copied circuit. This tool is also used to paste a circuit copied by using the **Cut Circuit** tool or the **Copy Circuit** tool. To paste a circuit, select the panel in which you want to copy the circuit and then choose the **Paste Circuit** tool from the toolbar available at the bottom of the **Circuit Manager**; the circuit in the clipboard will be pasted in the selected panel.

## Calculate Wires

This tool is used to calculate the size of the wire to be used in the selected circuit.

# TUTORIAL

## Tutorial 1

In this tutorial, you will create an electrical layout of an office, refer to Figure 7-33. You can also download the architectural file from *www.cadcim.com*. Path of the file is: *Textbooks > CAD/CAM > AutoCAD MEP > AutoCAD MEP 2014 for Designers > Tutorial*.     **(Expected time: 30 min)**

Examine the model to determine the equipment to be added and parameters of wiring.

***Figure 7-33*** *An electrical layout of the office*

***Table 7-1*** *Parameters of the devices*

| Device | Fan | 1200 Long Linear Floroscent | Single Switch Socket |
|---|---|---|---|
| Load | 180VA | 150VA | 300VA |
| Voltage | 230 | 230 | 230 |
| Phases | 1 | 1 | 1 |
| Maximum Overcurrent Rating | 10 | 8 | 10 |

The following steps are required to complete this tutorial:

a.   Open the drawing downloaded from the website.
b.   Add the devices according to the layout.
c.   Configure the devices to apply specific load.
d.   Create wiring line between devices.
e.   Calculate load and wire size of the circuits

## Downloading and Opening the Drawing File

1.   Download the c07_amep_prt.zip file from the *http://www.cadcim.com*. The path of the file is as follows: *Textbooks > CAD/CAM > AutoCAD MEP > AutoCAD MEP 2014 for Designers > Input*

2.   Extract this file to desired location.

3.   Open the drawing file *c07_amep_prt.dwg* from the specified location by double-clicking on it. The drawing file is displayed, as shown in Figure 7-34.

***Figure 7-34*** *Architectural layout of the office*

## Adding the Devices

1.  Open the **Electrical** workspace and choose the **Device** tool from the **Build** panel in the **Home** tab of the **Ribbon**; the **Properties** palette is displayed, as shown in Figure 7-35.

*Figure 7-35  The **Properties** palette displayed after choosing the **Device** tool*

2.  Click in the **Style** field in the **BASIC** rollout of the **Properties** palette; the **Select a Style** dialog box will be displayed, refer to Figure 7-36.

3.  Choose the **Lighting Fluorescent (Global)** option from the **Drawing file** drop-down list in the dialog box and select the **1200 Long Linear Fluorescent** device from the area displaying devices in the dialog box. Now, choose the **OK** button from the dialog box, the device gets attached to the cursor.

4.  Place the lights, as shown in Figure 7-37.

5.  Similarly, place sockets and fans in the drawing area, refer to Figure 7-38.

**Note**
*The display of the symbols can be different if AutoCAD MEP is started by selecting the icon different from the AutoCAD MEP 2014 - English (Global) icon.*

*Figure 7-36* The *Select a Style* dialog box

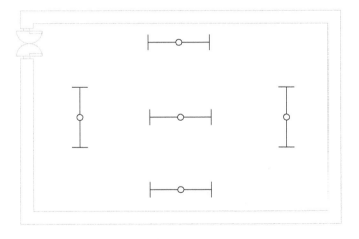

*Figure 7-37* The drawing after placing lights

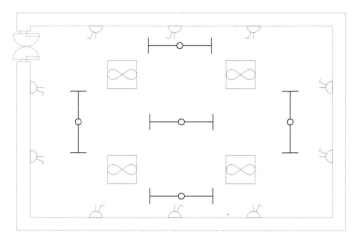

*Figure 7-38* *The drawing after placing fans and sockets*

Now, you need to place junction boxes in the drawing.

6.  Choose the **Device** tool from the **Build** panel in the **Home** tab of the **Ribbon**; the **Properties** palette is displayed.

7.  Click in the **Style** field in the **Properties** palette; the **Select a Style** dialog box will be displayed. Choose the **Junction Boxes (Global)** option from the **Drawing file** drop-down list; the junction boxes available in the current drawing are displayed.

8.  Select the **Hexagon 3 Junction Box** device from the devices displayed and then choose the **OK** button; the device gets attached to the cursor.

9.  Place the junction boxes, as shown in Figure 7-39.

## Configuring Devices

1.  Select a socket from the drawing area and then select the **Select Similar** option from the **Select Similar** drop-down list of the **Device** contextual tab displayed in the **Ribbon**; all the sockets available in the drawing area are selected.

2.  Click on the **Electrical properties** field in the **Circuit** rollout of **ADVANCED** rollout in the **Properties** palette; the **Electrical Properties** dialog box is displayed, as shown in Figure 7-40.

    Now, you need to specify the parameters as per the table.

3.  Select the **230V Power** option from the **System** drop-down list.

4.  Specify **300** in the **Load Phase 1** edit box, refer to Table 7-1.

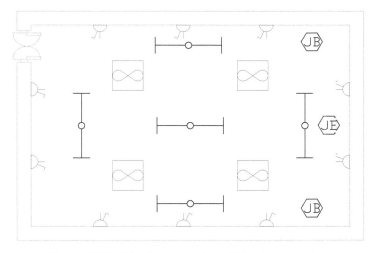

**Figure 7-39** *The drawing after adding junction boxes*

**Figure 7-40** *The **Electrical Properties** dialog box*

5. Select the **230** option from the **Voltage** drop-down list.

6. Similarly, select the **1** option from the **Number of Poles** drop-down list in the dialog box.

7. Specify **10** in the **Maximum Overcurrent Rating (amps)** edit box in the dialog box.

8. Specify **0.8** in the **Power Factor** edit box. Choose the **OK** button to exit the dialog box.

9. Similarly, specify the parameters for other devices according to the table in the **Electrical Properties** dialog box by following the steps discussed above. Note that the power factor will be same for all the devices.

## Creating Panel
All the devices are connected to circuits. These circuits are joined to a panel for supply. So, you will create a panel with circuits in this section.

1. Choose the **Panel** tool from the **Build** panel in the **Home** tab of the **Ribbon**; you are prompted to specify an insertion point for the panel and the **Properties** palette after choosing the **Panel** tool is displayed, as shown in Figure 7-41.

2. Specify the name as **Main Panel** in the **Name** edit box in the **Design Data** rollout of the **ADVANCED** rollout of the **Properties** palette.

3. Specify **800** in the **Rating** edit box.

4. Select the **230** option from the **Voltage phase-to-neutral** drop-down list.

5. Select the **240** option from the **Voltage phase-to-phase** drop-down list.

*Figure 7-41 The **Properties** palette displayed after choosing the **Panel** tool*

6. Specify main type as main circuit breaker, main size (amps) as **15**, design capacity (amps) as **20**, and AIC rating as **800** in the **Properties** palette.

Now, you need to create circuits for the panel.

7. Click in the **Circuit Settings** field in the **Properties** palette; the **Circuit Settings** dialog box is displayed, as shown in Figure 7-42.

*Figure 7-42  The Circuit Settings dialog box*

8. Make sure that **Power and Lighting** is selected in the **System Type** drop-down list and **230V Power** is selected in the **System** drop-down list. Set the value as **3** in both the **Total number of slots** and **Number of 1-pole circuits** spinners and select the **230** option from the **Voltage** drop-down lists adjacent to the spinners.

9. Choose the **OK** button from the dialog box to exit.

10. Click in the drawing area to place the panel, refer to Figure 7-43.

### Adding Wires

1. Choose the **Wire** tool from the **Build** panel in the **Home** tab of the **Ribbon**; you are prompted to specify the start point of the wire on an electrical device and the **Properties** palette is displayed, as shown in Figure 7-44.

2. Select **Spline** from the **Segment** drop-down list and click in the **System** drop-down list and select the **230V Power(230V POWER)** option from the drop-down list.

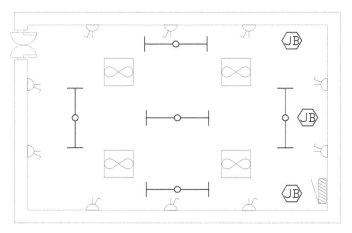

**Figure 7-43** *The drawing after adding the panel*

**Figure 7-44** *The **Properties** palette displayed on choosing the **Wire** tool*

3.   Connect all the sockets to the junction box using wires, refer to Figure 7-45. Make sure that the **Line** option is selected in the **Segment** drop-down list.

4.  Similarly, connect all the fans and lights to the junction boxes using the wires, refer to Figure 7-46. Make sure that system selected for fans is **230V Power** and for lights is **230V Lighting Devices (Ceiling)**.

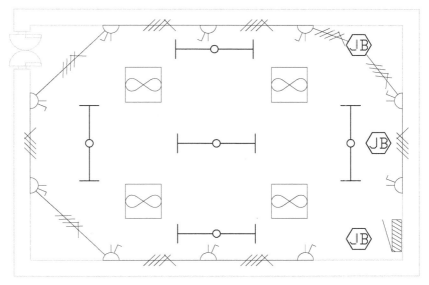

*Figure 7-45*  *The drawing after adding wires to the sockets*

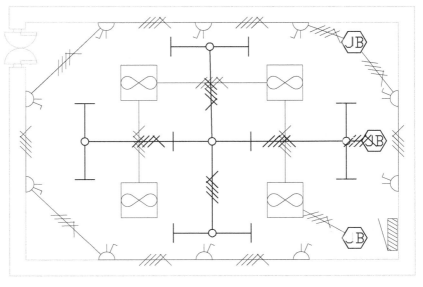

*Figure 7-46*  *The drawing after adding wires to lights and fans*

**Note**
*A device or wire related to **230V Power** system is displayed in cyan color and a device or wire related to **230V Lighting Devices (Ceiling)** is displayed in brown color in the drawing area. Make sure that all the devices connected to a circuit are in the same system.*

5.  Connect all the junction boxes to the panel, refer to Figure 7-47.

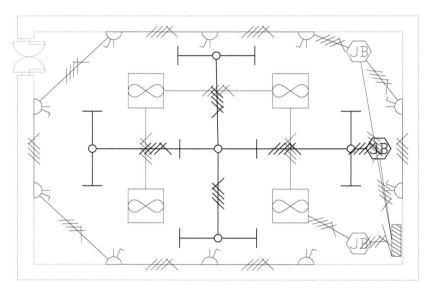

*Figure 7-47  The drawing after connecting junction boxes to the panel*

## Calculating Loads and Wire sizes

1.  To calculate the total load of all the devices in the drawing area, choose the **Power Totals** tool from the **Electrical** panel in the **Analyze** tab of the **Ribbon**; you are prompted to select the devices.

2.  Select all the devices available in the drawing area and press ENTER; the **Power Totals** dialog box is displayed, refer to Figure 7-48.

    The total load is displayed in the **Total Load** field of the dialog box. Now, you need to calculate the wire size for the circuit.

3.  To calculate wire size, select all the wires in the drawing area; the **Properties** palette is displayed.

4.  Click on the **Calculate sizes for the wire** button available at the right of the **Dimensions** rollout in the **ADVANCED** rollout of the **Properties** palette; the wire sizes are displayed in the **Hot size**, **Neutral size**, and **Ground size** edit boxes in the **Dimensions** rollout.

## Saving the drawing file

1.  Choose **Save** from the **Application** menu to save the drawing file.

*Figure 7-48  The **Power Totals** dialog box*

## Self-Evaluation Test

**Answer the following questions and then compare them to those given at the end of this chapter:**

1.  The **Create Circuits** option is available in the **Properties** palette displayed on selecting a wire. (T/F)

2.  The cable tray is used to support the cables in the system. (T/F)

3.  The **Create Circuits** option is available in the **Properties** palette displayed while creating a _____.

4.  The _____ is used to create and manage circuits in AutoCAD MEP.

5.  In which of the following tabs is the **Circuit Manager** tool available?

    (a) **Home**          (b) **Analyze**
    (c) **View**          (d) **Manage**

## Review Questions

**Answer the following questions:**

1.  The **Space** tool is not available in the **Electrical** workspace. (T/F)

2.  The **Hair Drier** option is not available in the **Select a style** dialog box. (T/F)

3.  You can select a wire type from the _____ tab of the **Tool Palette - Electrical**.

4.  The Circuit Report is used to display data related to various parameters of a _____.

5.  Which of the following tools has the **Cut Length** option?

    (a) **Conduit**          (b) **Cable Tray**
    (c) **Wire**          (d) **Panel**

# Exercise

## Exercise 1

In this exercise, you will create a model of an electrical system, refer to Figure 7-49. The architectural drawing for creating this exercise is available at *www.cadcim.com*. Download the c06_amep_prt.zip file from the *http://www.cadcim.com*. The path of the file is as follows: *Textbooks > CAD/CAM > AutoCAD MEP > AutoCAD MEP 2014 for Designers > Input*. Table 7-2 contains the parameters of the devices to be added to the layout.          **(Expected time: 30 min)**

*Figure 7-49* The final model

*Table 7-2* Parameters for the devices

| Device | Fan | 300x300 Recessed Light | Single Switch Socket |
|---|---|---|---|
| Load | 180VA | 100 | 300VA |
| Voltage | 230 | 230 | 230 |
| Phases | 1 | 1 | 1 |
| Maximum Overcurrent Rating | 10 | 8 | 10 |

## Answers to Self-Evaluation Test
**1.** F, **2.** T, **3.** panel, **4. Circuit Manager**, **5. Analyze**

# Chapter 8

## Representation and Schedules

*After completing this chapter, you will be able to:*

- *Create vertical section of the model*
- *Create horizontal section of the model*
- *Create section line for creating section views*
- *Create elevation line*
- *Create hidden line projections*
- *Slice a 3D object for extracting a profile*
- *Create hidden line projections*
- *Create schedules*
- *Create tables*

# INTRODUCTION

In this chapter, you will learn to create different views of the building model for representation. These views represent every minute detail of the building model. You will also learn to create various schedules and tables to document the equipment required for the building.

# CREATING VERTICAL SECTION

Vertical section is the section created by a vertical plane passing through the object to be sectioned. To create a section, choose the **Vertical Section** tool from the **Section & Elevation** panel of the **Home** tab in the **Ribbon**; you will be prompted to specify the start point of the section line. Click in the drawing area to specify the start point of the section; you will be prompted to specify the next point of the section line.

Click in the drawing area to specify the end point of section line; you will be prompted again to specify the next point for the section line. Press ENTER; you will be prompted to specify the length of the bounding box. Enter the length of the section area at the command prompt; the section line will be created. Figure 8-1 shows a drawing with the section line created. For section height, the software automatically assumes the height of section as the vertical extent of the building. You can change this default height to a user defined value. To do so, select the line and click in the **Use model extents for height** field in the **Properties** palette and select the **No** option from the drop-down list; the **Height** and **Lower extension** fields will become available below the **Use model extents for height** field. Click in the **Height** field and specify the value of section height. You can also specify the value of lower extension in the **Lower extension** field.

Section line created

***Figure 8-1*** *The drawing after creating the section line*

After specifying the section line, you will create a section view by using the section line. To create a vertical section view, select the section line; the **Building Section Line** contextual tab will be displayed, as shown in Figure 8-2. The options available in the contextual tab are discussed next.

*Figure 8-2* *The **Building Section Line** contextual tab*

# Enable Live Section

This tool is used to create live sections by using the current section line. This type of section is a dynamic section of the building and gets modified as you modify section line. To create a live section, select the section line and then choose the **Enable Live Section** tool from the **Live Section** panel of the **Building Section Line** contextual tab in the **Ribbon**; the live section will be created. To display the live section, you must switch to any of the isometric view such as SW Isometric and NE Isometric. Figure 8-3 shows a drawing with the live section created.

*Figure 8-3* *The drawing with live section created*

# Disable Live Section

This tool is used to remove the live sections created earlier. To remove a live section, select the section line that was used to create it; the **Building Section Line** contextual tab will be displayed. To remove the live section, choose the **Disable Live Section** tool from the **Live Section** panel of the **Building Section Line** contextual tab in the **Ribbon**; the live section will be removed.

# Toggle Body Display

This tool is used to toggle the display of object that is not displayed in the section.

# Reverse

This tool is used to reverse the side of the bounding box with respect to the section line. To reverse the side of the bounding box, choose the **Reverse** tool from the **Modify** panel in the **Building Section Line** contextual tab of the **Ribbon**; the bounding box will be created on the reverse side of the section line.

# Generate Section

This tool is used to generate section view of a building with respect to the section line selected. To generate a section, choose the **Generate Section** tool from the **Modify** panel of the **Building Section Line** contextual tab in the **Ribbon**; the **Generate Section/Elevation** dialog box will be displayed, refer to Figure 8-4. Choose the **Select Objects** button in the **Selection Set** area of the dialog box; you will be prompted to select the objects that are to be included in the section view. After selecting the object to be sectioned, press ENTER. Now, you need to define the type of section result. In AutoCAD MEP, you can display two type of section results by using their respective radio buttons from the Result Type area of the dialog box. Both these options of generating section views are discussed next.

*Figure 8-4  The **Generate Section/Elevation** dialog box*

## 3D Section/Elevation Object

This radio button is used for creating 3D section view of the building. If you select this radio button while creating the section view, the section will be created three dimensionally, refer to Figure 8-5.

*Figure 8-5* *The 3D section view of the building*

## 2D Section/Elevation Object with Hidden Line Removal

This radio button is used for creating 2D section view of the building. Figure 8-6 shows a 2D section created.

*Figure 8-6* *The drawing after creating the 2D section*

# CREATING HORIZONTAL SECTION

The **Horizontal Section** tool is used to create a section of the building bounded by two horizontal planes. To create a horizontal section, choose the **Horizontal Section** tool from the **Section & Elevation** panel of the **Home** tab in the **Ribbon**; you will be prompted to select a corner for horizontal section. Click in the drawing area to specify first corner of the section plane; you will be prompted to specify the other diagonal corner point of the section plane. Click to specify the point; you will be prompted to enter elevation of the current section plane. Enter the elevation value at the command prompt; you will be prompted to specify the depth of the section. Enter the desired value at the command prompt; the section lines will be created. To create a section, select the created section lines from the drawing area; the **Building Section Line** tab will be displayed in the **Ribbon**. Choose the **Generate Section** tool from the **Modify** panel in the contextual tab; the **Generate Section/Elevation** dialog box will be displayed. Create the section as discussed earlier. The options in the **Building Section Line** contextual tab have already been discussed.

# CREATING SECTION LINE

The **Section Line** tool is used to create section line passing through user defined points. To create a section line, choose the **Section Line** tool from the **Section & Elevation** panel of the **Home** tab in the **Ribbon**; you will be prompted to specify start point of the section line. Click in the drawing area to specify the start point of the section; you will be prompted to specify the next point of the section line. Click in the drawing area to specify the end point of section line; you will be prompted again to specify the next point for the section line. Press ENTER to exit creating section line; you will be prompted to specify length of the bounding box. Enter the length of the bounding box at the command prompt; the section line will be created. By default, model extent is used as the bounding box height. You can specify the height of the bounding box as discussed earlier. To create section view, select the section line; the **Building Section Line** contextual tab will be displayed in the **Ribbon**. Rest of the procedure to create section view is same as discussed earlier.

# CREATING ELEVATION LINE

The **Elevation Line** tool is used to create an elevation line passing through user defined points. To create an elevation line, choose the **Elevation Line** tool from the **Section & Elevation** panel of the **Home** tab in the **Ribbon**; you will be prompted to specify the start point of the elevation line. Click in the drawing area to specify the start point of the elevation line; you will be prompted to specify the end point of the elevation. Click in the drawing area to specify the end point of the elevation line; the elevation line and the elevation plane will be created. The plane created is square in shape. You can change the length of Side1 and Side2 by using the corresponding options in the **Properties** palette, refer to Figure 8-7. You can also change the angle values of the plane by using the **Angle1** and **Angle2** edit boxes.

Now, select the elevation line; the **Building Section Line** contextual tab is displayed in the **Ribbon**. Using the options available in the contextual tab, you can create the section as discussed earlier.

# CREATING HIDDEN LINE PROJECTION

The **Hidden Line Projection** tool is used to create 2D projections with hidden projection lines of the model. To create the projection, choose the **Hidden Line Projection** tool from the expanded **Section & Elevation** panel of the **Home** tab in the **Ribbon**; you will be prompted to select objects for creating a projection. Select all objects of the model to generate the 2D projections with hidden lines and then press and press ENTER; you will be prompted to specify the insertion point for the projection. Click in the drawing area to specify the insertion point; you will be prompted to specify whether the projection is to be inserted in plan view or not. Enter **Y** at the command prompt if you want to insert the plan view otherwise the isometric view will be inserted in the drawing. Figure 8-8 shows a model with its hidden line projection.

*Figure 8-7  The **Properties** palette*

*Figure 8-8  The model with its hidden line projection*

# SLICING THE MODEL

The **Quick Slice** tool is used to extract a polyline outline of the slice created through the model. To extract the outline, choose the **Quick Slice** tool from the expanded **Section & Elevation** panel in the **Home** tab of the **Ribbon**; you will be prompted to select the objects to be sliced. Select the objects to be sliced and press ENTER; you will be prompted to select first point for slicing. Click in the drawing area to specify the first point; you will be prompted to specify the next point for slicing. Click to specify the next point; the slice will be created through the specified points. Figure 8-9 shows a model sliced by using the **Quick Slice** tool.

*Figure 8-9* *The model sliced using the **Quick Slice** tool*

# REFRESHING SECTIONS AND ELEVATIONS IN A BATCH

The **Batch Refresh** tool is used to refresh all the 2D sections and elevations created in the current project or in the specified folder. To refresh the 2D sections and elevations, choose the **Batch Refresh** tool from the expanded **Section & Elevation** panel of the **Home** tab in the **Ribbon**; the **Batch Refresh 2D Section/Elevations** dialog box will be displayed, refer to Figure 8-10.

There are two radio buttons available in this dialog box: **Current Project** and **Folder**. Select the **Current Project** radio button if you want to refresh all the 2D sections and elevations created in the current project. If you want to refresh all the 2D sections and elevations in the drawings of a specific folder then select the **Folder** radio button. After selecting the desired radio button, choose the **Begin** button; all the section and elevation views available in the selected drawings will be updated and the status will be displayed in the **Status** area of the dialog box.

*Figure 8-10  The **Batch Refresh 2D Section/Elevations** dialog box*

# INSERTING DETAIL COMPONENTS

The detail components are used to represent AutoCAD MEP components with detailed parameters. To insert detail components, choose the **Detail Components** tool from the **Details** panel of the **Home** tab in the **Ribbon**; the **Detail Component Manager** dialog box will be displayed, refer to Figure 8-11. In the left area of this dialog box, various categories of the detail components are displayed. Click on the plus sign adjacent to the desired category in the left area; the sub-categories available in that category will be displayed. Click on the plus sign adjacent to the desired sub-category; various components in that sub-category will be displayed, refer to Figure 8-12. Select a component from the list; various sizes for the selected component will be displayed in the table available at the bottom of the dialog box. Select the desired size from the table; the **Insert Component** button will be activated. Choose the **Insert Component** button from the dialog box; you will be prompted to specify the insertion point for the component. Click on the desired location in the drawing area to specify the insertion point; you will be prompted again to specify the insertion point. Click to specify the insertion point for another component if required or press ENTER to exit the tool. You can change the view of the component to be inserted by using the options in the **View** drop-down list in the **Properties** palette.

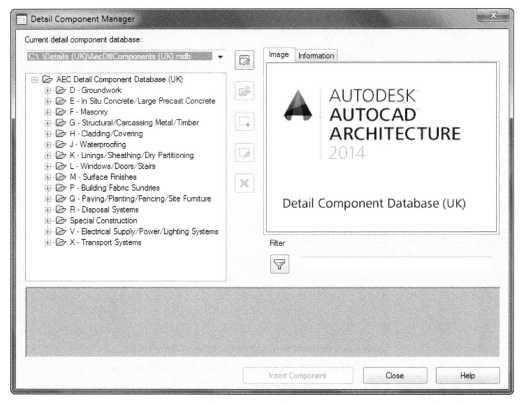

*Figure 8-11  The **Detail Component Manager** dialog box*

Some of the options available in the **Detail Component Manager** are discussed next.

## Edit Database

This button is available at the right of the **Current detail component database** drop-down list. Using this button, you can edit the database available for detail components. On choosing this button; the buttons available below it will be activated.

## Add Group

This button is available below the **Edit Database** button and is used to add groups to the database. To add groups, choose the **Add Group** button; the **Add Group** dialog box will be displayed, as shown in Figure 8-13. The options available in this dialog box are discussed next.

*Figure 8-12  The **Detail Component Manager** dialog box with the component selected*

*Figure 8-13  The **Add Group** dialog box*

## Group Name
This edit box is used to specify the name of the detail components group.

## Path Key

The options in this drop-down list are used to identify the location of images and drawings for this group. By default, the **Same as Parent Group** option is selected in this drop-down list so the path key is same as for the parent group.

## Filter Keywords

This edit box is used to specify the keywords for searching components in the current group.

# Add Component

This button is available below the **Add Group** button. This button is used to add components to the selected group. To add a component, choose the **Add Component** button; the **New Component** dialog box will be displayed, as shown in Figure 8-14. The options available in the **General** tab of this dialog box are discussed next.

*Figure 8-14  The New Component dialog box*

## Display Name

This edit box is used to specify the name of the component that will be displayed in the **Detail Component Manager** dialog box.

## Table Name

This edit box is used to specify the name of the size table for the current component.

## Recipe

This edit box is used to specify the name of the file that consists the method of creating the current component.

## Units

The options in this drop-down list are used to specify whether the current unit system is mm or inches.

## Description

This edit box is used to specify the description about the component being created.

## Keynote

This edit box is used to assign key notes about the component being created. You can also assign the key notes by selecting them from the **Select Keynote** dialog box. The **Select Keynote** dialog box will be displayed on choosing the **Select Keynote** button from the **Component Properties** dialog box, refer to Figure 8-15. Click on the plus sign adjacent to the desired category and then the sub-category; the keynotes for various components in that sub-category will be displayed. Select the desired keynote and then choose the **OK** button to add the selected keynote. On adding the keynote, the description about the selected keynote will be displayed in the adjacent display box.

## Filter Keywords

This edit box is used to specify keywords for which the created component will be displayed in the search list.

## Author/Manufacturer

This edit box is used to specify the name of the author/manufacturer of the current component.

## Web Address

This edit box is used to specify the website url of the current component to be created.

## Date Created

This display box is used to display the date on which the component was created.

*Figure 8-15*  *The **Select Keynote** dialog box*

## Date Modified

This display box is used to display the date on which the component was modified. If you are creating a new component, then the current date is displayed in this edit box.

## ID

This edit box is used to assign a unique identity number to the current component. By default, the system provides the unique number in this edit box.

After specifying the desired values in the **General** tab, click on the **Parameters** tab to display the options related to parameters of the current component, refer to Figure 8-16. The options available in the **Parameters** tab of the dialog box are discussed next.

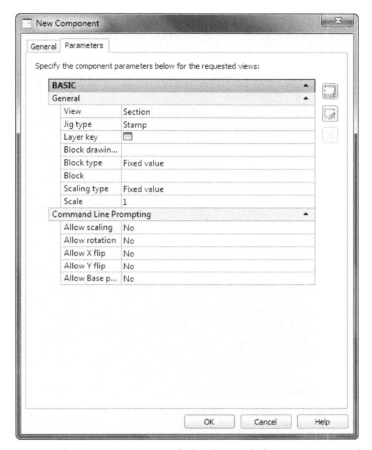

*Figure 8-16  The **New Component** dialog box with the **Parameters** tab selected*

## View

This field is used to specify the name of the view for which the current component will be available.

## Jig type

The options in this drop-down list are used to specify the pattern in which the current component will be inserted in the drawing. There are six options available in this drop-down list: **Stamp**, **Bookends**, **Linear Array**, **Surface**, **Surface Linetype**, and **Surface Top**.

## Layer Key

This field is used to specify a layer key for the current component. Click on this field; the **Select Layer Key** dialog box will be displayed, refer to Figure 8-17.

*Figure 8-17  The **Select Layer Key** dialog box*

Select a layer key from the **Layer Key** column of the dialog box and choose the **OK** button to exit the dialog box. The **Layer Key** option is also available in the **Hatching** and **Linetype** rollouts in the dialog box and has the same function. These rollouts are displayed on selecting various options from the **Jig type** drop-down list.

## Block drawing location
The options in this drop-down list are used to select the location of block drawing for the component. Select the **Browse** option from the drop-down list to select a block drawing file from a user defined location; the **Select Block Library** dialog box will be displayed, as shown in Figure 8-18. Select the block drawing and then choose the **Open** button from the dialog box; the selected drawing will be used as a block for the current component.

## Block type
The options in this drop-down list are used to select the type of block to be used for current component while inserting in the drawing. There are two options available in this drop-down list: **Fixed value** and **Database**. The **Fixed value** option is selected if the current block is fixed. The **Database** option is selected if you want to create the block of sizes dependent on the option selected in the database.

## Block
This edit box is used to specify a name for the current drawing block.

## Scaling type
The options in this drop-down list are used to specify the type of scaling to be used for scaling the current component. By default, the **Fixed value** option is selected in this drop-down list. You can select the **Database** option if you want to change the scale value depending on the option selected in the database.

*Figure 8-18* The **Select Block Library** *dialog box*

## Scale/Scale Field

These edit boxes are used to specify the value of scale. If the **Fixed value** option is selected in the **Scaling type** drop-down list then the **Scale** edit box will be displayed. You can specify an integral value in the **Scale** edit box. If the **Database** option is selected in the **Scaling Type** drop-down list, then the **Scale Field** edit box will be displayed. By default, the S_SCALE value is displayed in this edit box.

## Allow scaling

There are two options available in this drop-down list: **Yes** and **No**. If you select the **Yes** option in this drop-down list, you will be prompted to scale the component while inserting it in the drawing.

## Allow rotation

There are two options available in this drop-down list: **Yes** and **No**. If you select the **Yes** option in this drop-down list, you will be prompted to rotate the component while inserting it in the drawing.

## Allow X flip

There are two options available in this drop-down list: **Yes** and **No**. If you select the **Yes** option in this drop-down list, you will be prompted to flip the component about X axis while inserting it in the drawing.

## Allow Y flip
There are two options in available this drop-down list: **Yes** and **No**. If you select the **Yes** option in this drop-down list, you will be prompted to flip the component about Y axis while inserting in the drawing.

## Allow Base point
There are two options in available this drop-down list: **Yes** and **No**. If you select the **Yes** option in this drop-down list, you will be prompted to specify the base point for the component while inserting in the drawing.

## Create new view
This tool is available at the right of the dialog box. This tool is used to create more views of the component. To create more views, choose the **Create new view** button from the dialog box; the **New Component View** dialog box will be displayed, as shown in Figure 8-19.

*Figure 8-19  The **New Component View** dialog box*

Specify the name of the view in the edit box available in this dialog box and choose the **OK** button from the dialog box; a new component view with the specified name will be added to the **View** drop-down list.

## Rename view
This tool is available below the **Create new view** tool. This tool is used to rename the current selected view of the component. To rename the view, choose the **Rename view** button; the **Rename View** dialog box will be displayed, as shown in Figure 8-20. Specify the name of the view and then choose the **OK** button; the current view will be renamed.

*Figure 8-20  The **Rename View** dialog box*

## Delete view
This tool is available below the **Rename view** tool. This tool is used to delete the current selected view of the component. To delete the current view, choose the **Delete view** button; the **Delete Component View** dialog box will be displayed, as shown in Figure 8-21. Choose the **Yes** button if you want to delete the current view.

*Figure 8-21* The **Delete Component View** *dialog box*

After specifying the desired options, choose the **OK** button from the **New Component** dialog box; the component will be added to the selected category of components in the **Detail Component Manager** dialog box.

## Edit

This tool is available below the **Add Component** tool. It is used to edit current component. To edit the selected component, choose the **Edit** button; the **Component Properties** dialog box will be displayed, as shown in Figure 8-22. The options in this dialog box are similar to those discussed for the **New Component** dialog box.

*Figure 8-22* The **Component Properties** *dialog box*

## Delete

This button is used to delete the selected component from the **Detail Component Manager** dialog box. To delete a component, choose the **Delete** button; the **Confirm Component Delete** dialog box will be displayed, as shown in Figure 8-23. Choose the **Yes** button from the dialog box; the selected component will be deleted.

*Figure 8-23  The **Confirm Component Delete** dialog box*

# CREATING SCHEDULES

The schedules are the tables that are used to represent the information regarding the selected components. Some of the schedules available in AutoCAD MEP are: Plumbing Fixture Schedule, Air Terminal Devices Schedules, and Fan Schedules. The tools to create these schedules are discussed next.

## Air Terminal Devices Schedule

This tool is available in the **Schedules** drop-down in the **Annotation** panel of the **Home** tab of the **Ribbon** when the **HVAC** option is selected in the **Workspace Switching** flyout. To create an air terminal device schedule, choose the **Air Terminal Devices Schedule** tool from the drop-down; you will be prompted to select objects or press ENTER to schedule external drawings. Select the objects that you want in the schedule and then press ENTER; the schedule will get attached to the cursor and you will be prompted to specify the lower left corner of the table. Click in the drawing area to specify the upper left corner of the schedule; you will be prompted to specify the lower right corner of the schedule. Click in the drawing area to specify the lower right corner of the schedule; the table will be created, refer to 8-24. By default, the **?** mark is displayed in all the fields of the table. To update the entries in this schedule, select the schedule; the **Schedule Table** contextual tab will be added in the **Ribbon**, refer to Figure 8-25. Choose the **Add All Property Sets** tool from the **Modify** panel of the **Schedule Table** contextual tab in the **Ribbon**; the fields in the table will display the parameters related to the selected devices. The options available in the contextual tab are discussed next.

| MECHANICAL AIR TERMINAL DEVICES SCHEDULE | | | | | | | | |
|---|---|---|---|---|---|---|---|---|
| TAG | SIZE | DESCRIPTION | CONSTRUCTION | | | BASIS OF DESIGN | | NOTES |
| | | | MATERIAL | FINISH | DISCHARGE PATTERN | MANUFACTURER | MODEL OR SERIES | |
| ? | ? | ? | ? | ? | ? | ? | ? | ? |
| ? | ? | ? | ? | ? | ? | ? | ? | ? |
| ? | ? | ? | ? | ? | ? | ? | ? | ? |
| ? | ? | ? | ? | ? | ? | ? | ? | ? |
| ? | ? | ? | ? | ? | ? | ? | ? | ? |
| ? | ? | ? | ? | ? | ? | ? | ? | ? |
| ? | ? | ? | ? | ? | ? | ? | ? | ? |
| ? | ? | ? | ? | ? | ? | ? | ? | ? |
| ? | ? | ? | ? | ? | ? | ? | ? | ? |
| ? | ? | ? | ? | ? | ? | ? | ? | ? |
| ? | ? | ? | ? | ? | ? | ? | ? | ? |
| ? | ? | ? | ? | ? | ? | ? | ? | ? |
| ? | ? | ? | ? | ? | ? | ? | ? | ? |
| ? | ? | ? | ? | ? | ? | ? | ? | ? |
| ? | ? | ? | ? | ? | ? | ? | ? | ? |
| ? | ? | ? | ? | ? | ? | ? | ? | ? |
| ? | ? | ? | ? | ? | ? | ? | ? | ? |
| ? | ? | ? | ? | ? | ? | ? | ? | ? |
| ? | ? | ? | ? | ? | ? | ? | ? | ? |
| ? | ? | ? | ? | ? | ? | ? | ? | ? |
| ? | ? | ? | ? | ? | ? | ? | ? | ? |

*Figure 8-24* The air terminal devices schedule

*Figure 8-25* The **Schedule Table** contextual tab

## General Panel

The tools in this panel are used to modify the general parameters of a schedule. The options in this panel are discussed next.

### Select Similar

This tool is used to select all the items in the current drawing similar to the selected one.

### Isolate Objects

The tools in this drop-down are used to display/hide objects in the drawing area and are discussed next.

#### Isolate Objects

This tool is used to display only the objects that are selected in the drawing. This tool is available in the **Isolate Objects** drop-down list.

#### Hide Objects

This tool is used to hide the selected object in the drawing. This tool is available in the **Isolate Objects** drop-down list.

#### End Isolation

This tool is used to end the isolation in the drawing. This tool is available in the **Isolate Objects** drop-down list.

**Edit Style**

The tools in this drop-down are used to edit styles and definitions of the schedule tables. You can also change the text styles and property data formats used in the schedule tables. The tools available in this drop-down are discussed next.

### Edit Style

This tool is used to change the style of the selected table. To change the style of the selected schedule, choose the **Edit Style** tool from the **Edit Style** drop-down; the **Schedule Table Style Properties** dialog box will be displayed, refer to Figure 8-26. The options in this dialog box are discussed next.

**General:** The options in this tab are used to specify the name and description of the current table style.

**Default Format:** The options in this tab are used to specify format such as text style, text height, text alignment, and cell size of the table.

**Applies To:** The options in this tab are used to select the categories for which the current selected style will be applied.

**Columns:** The options in this tab are used to add or delete columns in the table.

**Sorting/Grouping:** The options in this tab are used to sort or group the values specified in the table.

*Figure 8-26  The Schedule Table Style Properties - Duct Quantity dialog box*

**Layout:** The options in this tab are used to change layout of the table. Using these options, you can change the format of the title, column header, rows header and so on.

**Classifications:** The options in this tab are used to classify the values in the table.

**Display Properties:** The options in this tab are used to edit the display style of the table. Using these options, you can change properties such as colors and layers.

**Version History:** The options in this tab are used to display or edit the version history of the current table.

### Schedule Table Style

This tool is used to change the style of the selected schedule table. To change the style of the selected schedule table, choose the **Schedule Table Styles** tool from the **Edit Style** drop-down; the **Style Manager** dialog box will be displayed, refer to Figure 8-27.

The options in the **Style Manager** have already been discussed in the **Schedule Table Style Properties** dialog box.

*Figure 8-27  The Style Manager displayed on choosing the Schedule Table Styles tool*

### Classification Definition

This tool is used to change the classification style used in the current table. To change the classification style, choose the **Classification Definitions** tool from the **Edit Style** drop-down; the **Style Manager** will be displayed, as shown in Figure 8-28. Select a type from the left pane of the **Style Manager** and the related options will be displayed in the right of the dialog box. Using these options, you can change the style of classification of the components in the table.

*Figure 8-28  The Style Manager on choosing the Classification Definitions tool*

**Property Set Definitions**
This tool is used to change the definitions of property set for the selected property. To change the definition of property sets, choose the **Property Set Definitions** tool from the **Edit Style** drop-down; the **Style Manager** will be displayed, as shown in Figure 8-29. On selecting a part from the left pane, the options related to the selected part are displayed in the right area of dialog box. Using these options, you can change the definition of the property sets used in the schedule table.

**Property Data Formats**
This tool is used to change the format data specified in the schedule table. To change the data format, choose the **Property Data Formats** tool from the **Edit Style** drop-down; the **Style Manager** dialog box will be displayed, as shown in Figure 8-30. Using the options available in this dialog box, you can change the format of the properties.

*Figure 8-29* The **Style Manager** *on choosing the* **Property Set Definitions** *tool*

*Figure 8-30* The **Style Manager** *on choosing the* **Property Data Formats** *tool*

### Text Styles

This tool is used to change the text style in the schedule table. To change the text style, choose the **Text Style** tool from the **Edit Style** drop-down; the **Text Style** dialog box will be displayed, as shown in Figure 8-31. Using the options in this dialog box, you can change the style of text in the table.

### Copy Style

This tool is used to copy style of an existing object and then create a new style based on that style. To copy a style, choose the **Copy Style** tool; the **Schedule Table Style Properties** dialog box, as shown in Figure 8-32. The options in this dialog box have already been discussed.

*Figure 8-31*  *The* **Text Style** *dialog box*

## Modify panel

The tools in this panel are used to modify the properties of a schedule. The tools in this panel are discussed next.

### Update

This tool is used to update the fields in the table that are not up to date.

*Figure 8-32* The **Schedule Table Style Properties** *dialog box*

### Edit Table Cell

This tool is used to edit a table cell in the schedule. To do so, choose the **Edit Table Cell** tool; you will be prompted to select a schedule table item. Select an item from the table; the **Edit Referenced Property Set Data** dialog box will be displayed, refer to Figure 8-33. Using the fields available in this dialog box, you can change the properties of a table cell.

*Figure 8-33* The **Edit Referenced Property Set Data** *dialog box*

### Add All Property Sets

This tool is used to add all property sets in the table for which ? mark will be displayed in the table. To add the property sets, choose the **Add All Property Sets** tool from the contextual tab; the property sets in the table will be displayed automatically.

### Export

This tool is used to export the selected table in external formats. To export a table, select the table and then choose the **Export** tool; the **Export Schedule Table** dialog box will be displayed, as shown in Figure 8-34. Choose the desired file format from the **Save As Type** drop-down list in the **Export Schedule Table** dialog box and then choose the **OK** button; the table will be exported to the format selected in the **Save As Type** drop-down list. To change the location of the file to be exported, choose the **Browse** button and specify the desired location.

*Figure 8-34* The **Export Schedule Table** dialog box

### Convert to Table

This tool is used to convert a schedule into a table which can be edited directly. To convert a schedule into a table, select the schedule and then choose the **Convert to Table** tool from the **Modify** panel in the **Schedule Table** contextual tab; you will be prompted to specify an insertion point for the table. Click in the drawing area to specify the insertion point; the table will be placed at the specified point. To edit any of the cell in the table, select the cell; the **Table Cell** contextual tab will be added in the **Ribbon**. Using the options available in this tab, you can edit the table cells.

## Scheduled Objects

The tools in this panel are used to modify the objects in the schedule. These tools are discussed next.

### Add

This tool is used to add an object into the selected schedule table. To do so, choose the **Add** tool; you will be prompted to select the objects to be added in the table. Select the objects that you want to add in the schedule table and then press ENTER; the selected objects will be added in the schedule table.

### Remove

This tool is used to remove an object from the selected schedule table. To do so, choose the **Remove** tool; you will be prompted to select the objects to be removed from the table. Select the objects that you want to remove from the schedule table and then press ENTER; the selected objects will be removed from the schedule table.

### Reselect

This tool is used to clear all entries in the selected schedule table. To do so, choose the **Reselect** tool; you will be prompted to select the new objects that you want to add in the table. Select the objects and then press ENTER; the selected objects will be added to the schedule table and all the previous entries will be removed.

### Show

This tool is used to show the selected entity from the schedule table in the drawing area. To do so, choose the **Show** tool; you will be prompted to select schedule table entity. Select the entity; the selected entity will be highlighted in the drawing area.

## Fan Schedule

This tool is available in the **Schedules** drop-down in the **Annotation** panel of the **Home** tab of the **Ribbon** when the **HVAC** option is selected in the **Workspace Switching** flyout. To create a fan schedule, choose the **Fan Schedule** tool; you will be prompted to select the objects. The procedure for creating a fan schedule is similar to the procedure for creating an air terminal devices schedule.

## VAV Fan Powered Box (Electric Heat) Schedule

This tool is available in the **Schedules** drop-down in the **Annotation** panel of the **Home** tab in the **Ribbon** when the **HVAC** option is selected in the **Workspace Switching** flyout. To create a VAV fan powered box (electric heat) schedule, choose the **VAV Fan Powered Box (Electric Heat) Schedule** tool; you will be prompted to select the objects. The procedure for creating this type of schedule is similar to the procedure for creating an air terminal devices schedule.

## Space Engineering Schedule

This tool is available in the **Schedules** drop-down in the **Annotation** panel of the **Home** tab in the **Ribbon** when the **HVAC** option is selected in the **Workspace Switching** flyout. To create a space engineering schedule, choose the **Space Engineering Schedule** tool; you will be prompted to select the objects. The procedure for creating a space engineering schedule is similar to the procedure for creating an air terminal devices schedule.

## Duct Quantity Schedule

This tool is available in the **Schedules** drop-down in the **Annotation** panel of the **Home** tab in the **Ribbon** when the **HVAC** option is selected in the **Workspace Switching** flyout. To create a duct quantity schedule, choose the **Duct Quantity Schedule** tool; you will be prompted to select the objects. The procedure for creating a duct quantity schedule is similar to the procedure for creating an air terminal devices schedule.

## Duct Fabrication Contract Schedule

This tool is available in the **Schedules** drop-down in the **Annotation** panel of the **Home** tab in the **Ribbon** when the **HVAC** option is selected in the **Workspace Switching** flyout. To create a duct fabrication contract schedule, choose the **Duct Fabrication Contract Schedule** tool; you will be prompted to select the objects. The procedure for creating a duct quantity schedule is similar to the procedure for creating an air terminal devices schedule.

You can create user defined schedule using the **Table** tool. This tool is discussed next.

## Table

The **Table** tool is available in the all workspaces. To create a table, choose this tool from the **Schedule** drop-down in the **Annotation** panel of the **Home** tab in the **Ribbon**; the **Insert Table** dialog box will be displayed, as shown in Figure 8-35. The options in this dialog box are discussed next.

*Figure 8-35  The **Insert Table** dialog box*

## Table style

The options in the **Table style** drop-down list are used to select a table style for the current table. You can edit the selected table style by using the options available in the **Table Style** dialog box. To invoke the **Table Style** dialog box, choose the **Launch the Table Style dialog** button adjacent to the **Table style** drop-down list; the **Table Style** dialog box will be displayed, as shown in Figure 8-36.

*Figure 8-36  The Table Style dialog box*

## Insert options

The radio buttons in this area are used to specify the insertion method of the table. There are three radio buttons available in this area: **Start from empty table**, **From a data link**, and **From object data in the drawing (Data Extraction)**. If the **Start from empty table** radio button is selected then the table inserted will be empty. If you select the **From a data link** radio button then the table inserted will have the data linked to an excel sheet created earlier. If the **From object data in the drawing (Data Extraction)** radio button is selected then the table inserted consists of data extracted by using the **Data Extraction** dialog box.

## Insertion behavior

There are two radio buttons available in this area: **Specify insertion point** and **Specify window**. If the **Specify insertion point** radio button is selected then you will be prompted to specify the insertion point for the table. If you select the **Specify window** radio button from the **Insertion behavior** area then you will be prompted to create a window for inserting the table.

## Column & row settings

The options in this area are used to specify the number of columns and rows for the table. You can also specify the width of the column, the number of rows, and the height of the rows by using the related spinners.

## Set cell styles

The options in this area are used to specify the styles for cells. By default, the style of first row cell is set for Title, cell style of second row is set for Header and the cell style of third row is set for Data.

# Pipe & Fitting Schedule

This tool is available in the **Schedules** drop-down in the **Annotation** panel of the **Home** tab in the **Ribbon** when the **Piping** option is selected in the **Workspace Switching** flyout. To create a pipe and fitting schedule, choose the **Pipe & Fitting Schedule** tool; you will be prompted to select the objects. Select the pipes and fittings. The procedure for creating a pipe and fitting schedule is similar to the procedure for creating an air terminal devices schedule.

## Pipe Quantity

This tool is available in the **Schedules** drop-down in the **Annotation** panel of the **Home** tab in the **Ribbon** when the **Piping** option is selected in the **Workspace Switching** flyout. To create a pipe quantity schedule, choose the **Pipe Quantity** tool; you will be prompted to select the objects. Select the pipes in the drawing area. The procedure for creating a pipe quantity schedule is similar to the procedure for creating an air terminal devices schedule.

## Mechanical Pump Schedule

This tool is available in the **Schedules** drop-down in the **Annotation** panel of the **Home** tab of the **Ribbon** when the **Piping** option is selected in the **Workspace Switching** flyout. To create a mechanical pump schedule, choose the **Mechanical Pump Schedule** tool; you will be prompted to select the objects. Select the pumps from the drawing area. The procedure for creating a mechanical pump schedule is similar to the procedure for creating an air terminal devices schedule.

## Mechanical Tank Schedule

This tool is available in the **Schedules** drop-down in the **Annotation** panel of the **Home** tab in the **Ribbon** when the **Piping** option is selected in the **Workspace Switching** flyout. To create a mechanical tank schedule, choose the **Mechanical Tank Schedule** tool; you will be prompted to select the objects. Select the tanks in the drawing area. The procedure for creating a mechanical tank schedule is similar to the procedure for creating an air terminal devices schedule.

## Device Schedule

This tool is available in the **Schedules** drop-down in the **Annotation** panel of the **Home** tab in the **Ribbon** when the **Electrical** option is selected in the **Workspace Switching** flyout. To create a device schedule, choose the **Device Schedule** tool; you will be prompted to select the objects. The procedure for creating a device schedule is similar to the procedure for creating an air terminal devices schedule.

## Lighting Device Schedule

This tool is available in the **Schedules** drop-down in the **Annotation** panel of the **Home** tab in the **Ribbon** when the **Electrical** option is selected in the **Workspace Switching** flyout. To create a lighting device schedule, choose the **Lighting Device Schedule** tool; you will be prompted to select the objects. The procedure for creating a lighting device schedule is similar to the procedure for creating an air terminal devices schedule.

## Conduit & Fitting Schedule

This tool is available in the **Schedules** drop-down in the **Annotation** panel of the **Home** tab in the **Ribbon** when the **Electrical** option is selected in the **Workspace Switching** flyout. To create a conduit and fitting schedule, choose the **Conduit & Fitting Schedule** tool; you will be prompted to select the objects. The procedure for creating a conduit and fitting schedule is similar to the procedure for creating an air terminal devices schedule.

## Electrical & Mechanical Equipment Schedule

This tool is available in the **Schedules** drop-down in the **Annotation** panel of the **Home** tab in the **Ribbon** when the **Electrical** option is selected in the **Workspace Switching** flyout. To create an electrical and mechanical equipment schedule, choose the **Electrical & Mechanical Equipment Schedule** tool; you will be prompted to select the objects. The procedure for creating an electrical and mechanical equipment schedule is similar to the procedure for creating an air terminal devices schedule.

## 3-Phase Branch Panel Schedule

This tool is available in the **Schedules** drop-down in the **Annotation** panel of the **Home** tab in the **Ribbon** when the **Electrical** option is selected in the **Workspace Switching** flyout. To create a 3-phase branch panel schedule, choose the **3-Phase Branch Panel Schedule** tool; the **Panel Schedule** dialog box will be displayed, as shown in Figure 8-37. The options in this dialog box are discussed next.

*Figure 8-37  The **Panel Schedule** dialog box*

### Panel

The options in this drop-down list are used to select the panel for which the panel schedule is being created.

### Panel schedule table style

The options in this drop-down list are used to specify the style of panel schedule. There are two options available in this drop-down list: **Distribution Board** and **Panel**. Select the **Distribution Board** option if you want to use the distribution board as the panel schedule table style. If you want to use panel as panel schedule table style then select the **Panel** option from the drop-down list.

### Panel schedule style location

The options in this drop-down list are used to specify the location of panel schedule style. To specify a user defined location, select the **Browse** option from the drop-down list; the **Select**

**a file** dialog box will be displayed, as shown in Figure 8-38. Select the file and then choose the **Open** button; the panel schedule styles will be fetched from the selected file.

*Figure 8-38  The Select a File dialog box*

## Show panels from

There are two radio buttons in this area: **Current drawing** and **Electrical project database**. Select the **Current drawing** radio button if you want to use the panels available in the current drawing. You can also use the electrical project database for displaying panel. To do so, select the **Electrical project database** radio button; the system will automatically display the panels available in the electrical project database. After specifying the desired parameters in the dialog box, choose the **OK** button; the panel schedule will get attached to the cursor. Click in the drawing area to specify the insertion point; the schedule will be placed at the specified point. You can edit any of the cells in the schedule by double clicking on it.

## 1-Phase Branch Panel Schedule

This tool is also available in the **Schedules** drop-down in the **Annotation** panel of the **Home** tab in the **Ribbon** when the **Electrical** option is selected in the **Workspace Switching** flyout. To create a 1-phase branch panel schedule, choose the **1-Phase Branch Panel Schedule** tool; the **Panel Schedule** dialog box will be displayed. The options in this dialog box have already been discussed.

## Distribution Board Schedule

This tool is also available in the **Schedules** drop-down in the **Annotation** panel of the **Home** tab in the **Ribbon** when the **Electrical** option is selected in the **Workspace Switching** flyout. To create a distribution board schedule, choose the **Distribution Board Schedule** tool; the **Panel Schedule** dialog box will be displayed. The options in this dialog box have already been discussed.

## Switchboard Schedule

This tool is also available in the **Schedules** drop-down in the **Annotation** panel of the **Home** tab in the **Ribbon** when the **Electrical** option is selected in the **Workspace Switching** flyout. To create a switchboard schedule, choose the **Switchboard Schedule** tool; the **Panel Schedule** dialog box will be displayed. The options in this dialog box have already been discussed.

## Panel Schedule

This tool is also available in the **Schedules** drop-down in the **Annotation** panel of the **Home** tab in the **Ribbon** when the **Electrical** option is selected in the **Workspace Switching** flyout. To create a panel schedule, choose the **Panel Schedule** tool; the **Panel Schedule** dialog box will be displayed. The options in this dialog box have already been discussed.

## Plumbing Fixture Schedule

This tool is available in the **Schedules** drop-down in the **Annotation** panel of the **Home** tab in the **Ribbon** when the **Plumbing** option is selected in the **Workspace Switching** flyout. To create a plumbing fixture schedule, choose the **Plumbing Fixture Schedule** tool; you will be prompted to select the objects. The procedure for creating a plumbing fixture schedule is similar to the procedure for creating an air terminal devices schedule.

## Plumbing Fixture & Pipe Connection Schedule

This tool is available in the **Schedules** drop-down in the **Annotation** panel of the **Home** tab in the **Ribbon** when the **Plumbing** option is selected in the **Workspace Switching** flyout. To create a plumbing fixture and pipe connection schedule, choose the **Plumbing Fixture & Pipe Connection Schedule** tool; you will be prompted to select the objects. The procedure for creating a plumbing fixture and pipe connection schedule is similar to the procedure for creating an air terminal devices schedule.

## Water Heater (Gas) Schedule

This tool is available in the **Schedules** drop-down in the **Annotation** panel of the **Home** tab in the **Ribbon** when the **Plumbing** option is selected in the **Workspace Switching** flyout. To create a water heater schedule, choose the **Water Heater (Gas) Schedule** tool; you will be prompted to select the objects. The procedure for creating a water heater (gas) schedule is similar to the procedure for creating an air terminal devices schedule.

## Door Schedule

This tool is available in the **Schedules** drop-down in the **Annotation** panel of the **Home** tab in the **Ribbon** when the **Architecture** option is selected in the **Workspace Switching** flyout.

To create a door schedule, choose the **Door Schedule** tool; you will be prompted to select the objects. The procedure for creating a door schedule is similar to the procedure for creating an air terminal devices schedule.

## Door Schedule - Project Based

This tool is available in the **Schedules** drop-down in the **Annotation** panel of the **Home** tab in the **Ribbon** when the **Architecture** option is selected in the **Workspace Switching** flyout. To create a project based door schedule, choose the **Door Schedule - Project Based** tool; you will be prompted to select the objects. The procedure for creating a project based door schedule is similar to the procedure for creating an air terminal devices schedule.

## Window Schedule

This tool is available in the **Schedules** drop-down in the **Annotation** panel of the **Home** tab in the **Ribbon** when the **Architecture** option is selected in the **Workspace Switching** flyout. To create a window schedule, choose the **Window Schedule** tool; you will be prompted to select the objects. The procedure for creating a window schedule is similar to the procedure for creating an air terminal devices schedule.

## Room Schedule

This tool is available in the **Schedules** drop-down in the **Annotation** panel of the **Home** tab in the **Ribbon** when the **Architecture** option is selected in the **Workspace Switching** flyout. To create a room schedule, choose the **Window Schedule** tool; you will be prompted to select the objects. The procedure for creating a room schedule is similar to the procedure for creating an air terminal devices schedule.

## Space Schedule - BOMA

This tool is available in the **Schedules** drop-down in the **Annotation** panel of the **Home** tab in the **Ribbon** when the **Architecture** option is selected in the **Workspace Switching** flyout. To create a space schedule, choose the **Space Schedule - BOMA** tool; you will be prompted to select the objects. The procedure for creating a space schedule is similar to the procedure for creating an air terminal devices schedule.

## Space Inventory Schedule

This tool is available in the **Schedules** drop-down in the **Annotation** panel of the **Home** tab in the **Ribbon** when the **Architecture** option is selected in the **Workspace Switching** flyout. To create a space inventory schedule, choose the **Space Inventory Schedule** tool; you will be prompted to select the objects. The procedure for creating a space inventory schedule is similar to the procedure for creating an air terminal devices schedule.

## Wall Schedule

This tool is available in the **Schedules** drop-down in the **Annotation** panel of the **Home** tab in the **Ribbon** when the **Architecture** option is selected in the **Workspace Switching** flyout. To create a wall schedule, choose the **Wall Schedule** tool; you will be prompted to select the objects. The procedure for creating a wall schedule is similar to the procedure for creating an air terminal devices schedule.

## Schedule Styles

This tool is available in the **Schedules** drop-down in the **Annotation** panel of the **Home** tab in the **Ribbon** when the **Architecture** option is selected in the **Workspace Switching** flyout. This tool is used to open the **Style Manager** with the options related to schedule styles only. The options in the **Style Manager** have already been discussed.

## Table Editing

You can edit a schedule table in the same way as done in an excel sheet. On selecting a cell from the table, the **Table Cell** contextual tab will be displayed in the **Ribbon**, refer to Figure 8-39. The options in this tab are discussed next.

*Figure 8-39  The **Table Cell** contextual tab*

### Rows

The tools in this panel are used to add or delete rows from the table. The tools available in this panel are discussed next.

#### Insert Above

This tool is used to insert a row above the selected cell.

#### Insert Below

This tool is used to insert a row below the selected cell.

#### Delete Row(s)

This tool is used to delete the row of the selected cell.

### Columns

The tools in this panel are used to add or delete columns from the table. The tools available in this panel are discussed next.

#### Insert Left

This tool is used to insert a column on the left of the selected cell.

#### Insert Right

This tool is used to insert a column on the right of the selected cell.

#### Delete Column(s)

This tool is used to delete the column of the selected cell.

### Merge

The tools in this panel are used to merge or separate the cells in the table. The tools available in this panel are discussed next.

**Merge Cells**

The tools in this drop-down are used to merge the cells in the desired pattern. The tools in the drop-down are discussed next.

### Merge All

This tool is used to merge all the selected cells of the table.

### Merge By Row

This tool is used to merge all the selected cells row wise.

### Merge By Column

This tool is used to merge all the selected cells column wise.

**Unmerge Cells**

This tool is used to separate the merged cell in the table.

## Cell Style

The tools in this panel are used to manage cell styles. These tools are discussed next.

**Match Cell**

This tool is used to match the properties of other cells to the select cell. To use this option, select a cell and then choose the **Match Cell** tool; you will be prompted to select the destination cell. Click in the destination cell; the properties of the selected cell will be applied to the destination cell.

**Middle Center**

The tools available in this drop-down are used to specify the justification of text in the table cells. The tools available in this drop-down are discussed next.

### Top Left

It is used to align the text in the top left of the cell.

### Top Center

It is used to align the text in the top center of the cell.

### Top Right

It is used to align the text in the top right of the cell.

### Middle Left

It is used to align the text in the middle left of the cell.

### Middle Center

It is used to align the text in the middle center of the cell.

### Middle Right

It is used to align the text in the middle right of the cell.

**Bottom Left**
It is used to align the text in the bottom left of the cell.

**Bottom Center**
It is used to align the text in the bottom center of the cell.

**Bottom Right**
It is used to align the text in the bottom right of the cell.

### Table Cell Styles
The options in this drop-down list are used to apply a cell style to the selected cell. The cell styles displayed in this drop-down list are the cell styles available in the current table style.

### Create new cell style
This option is used to create new cell styles. To do so, select the **Create new cell style** option from the drop-down list; the **Create New Cell Style** dialog box will be displayed, refer to Figure 8-40. A default name is displayed in the **New Style Name** edit box of this dialog box. You can specify the desired name in this edit box. The options in the **Start With** drop-down list are used to select a template for the cell style to be created.

*Figure 8-40* *The* ***Create New Cell Style*** *dialog box*

After specifying the desired options, choose the **Continue** button; a new style will be created and displayed in the **Table Cell Styles** drop-down list.

### Table Cell Background Color
The options in this drop-down list are used to change the background color of the selected cell.

### Edit Border
This button is used to modify borders of the selected cell. To do so, choose the **Edit Border** button; the **Cell Border Properties** dialog box will be displayed, as shown in Figure 8-41. The options to edit the borders are divided into two areas: **Border properties** and **Border Type**. Using the options available in the **Border properties** area, you can change the properties of borders like line weight, line type, color and so on. To change the border type, choose the corresponding button from the **Border Type** area. Preview of the changes made in the selected border are displayed in the preview area available at the bottom of the dialog box. After specifying the desired options, choose the **OK** button from the dialog box; the changes will be applied to the selected cell.

*Figure 8-41  The **Cell Border Properties** dialog box*

## Cell Format

The tools in this panel are used to modify the format of data to be entered in a cell. Also, you can lock/unlock the selected cells by using the tools in this panel. These tools are discussed next.

### Cell Locking

The tools in this drop-down are used to lock/unlock the format, content, or both of a cell. The tools in this drop-down are discussed next.

#### Unlocked

This tool is used to unlock the selected cell so that it can be edited format wise as well as content wise.

#### Content Locked

This tool is used to lock the content of the selected cell so that it cannot be edited.

#### Format Locked

This tool is used to lock the format of the selected cell so that it cannot be edited.

### Content and Format Locked

This tool is used to lock the content as well as the format of the selected cell so that it cannot be edited.

### Data Format

The options available in this drop-down list are used to specify the format of the selected cell.

## Insert

The tools available in this panel are used to insert various types of objects in the table. These tools are discussed next.

### Block

This tool is used to insert a block in the current cell. To do so, choose the **Block** tool; the **Insert a Block in a Table Cell** dialog box will be displayed, refer to Figure 8-42. Using the options available in this dialog box; you can specify the parameters such as scale, rotation angle, and alignment for the selected block in the cell. A preview of the block to be inserted is displayed on the right in the dialog box. After specifying the desired options in this dialog box, choose the **OK** button; the block will be inserted in the selected cell.

*Figure 8-42* The **Insert a Block in a Table Cell** dialog box

### Field

This tool is used to insert a text field in the current cell. To insert a text field, choose the **Field** option; the **Field** dialog box will be displayed, as shown in Figure 8-43. Select a category from the **Field category** drop-down list in the dialog box; the options related to the selected category are displayed in the **Field names** list box. Select an option from this list box; the related options will be displayed in the right of the dialog box. Specify the desired option from the right area and then choose the **OK** button; the specified field will be inserted in the current cell.

*Figure 8-43  The **Field** dialog box*

### Formula
The tools available in this drop-down are used to apply a formula to the selected cell.

### Manage Cell Contents
This tool is used to manage the direction and order of the content in the selected cell of the table. This tool is active only when you insert a block in the cell.

## Data
The tools in this panel are used to manage data links of the selected cell. The tools available in this panel are discussed next.

### Link Cell
This tool is used to link the selected cell in the table to a field in the excel sheet. To do so, choose the **Link Cell** tool from the contextual tab; the **Select a Data Link** dialog box will be displayed, refer to Figure 8-44. Click on the **Create a new Excel Data Link** node in the **Links** area of the dialog box; the **Enter Data Link Name** dialog box will be displayed, as shown in Figure 8-45. Specify the name of the data link in the **Name** edit box and then choose the **OK** button from the dialog box; the **New Excel Data Link** dialog box will be displayed, as shown in Figure 8-46. Click on the Browse button next to the **Browse for a file** drop-down list in the dialog box; the **Save As** dialog box will be displayed and you

will be prompted to open an already existing excel file. Select an already existing excel file and then choose the **Open** button from the dialog box; the expanded **New Excel Data Link** dialog box will be displayed, refer to Figure 8-47.

*Figure 8-44* The *Select a Data Link* dialog box

*Figure 8-45* The *Enter Data Link Name* dialog box

*Figure 8-46* The *New Excel Data Link* dialog box

*Figure 8-47  The expanded **New Excel Data Link** dialog box*

This dialog box is divided into five areas: **File**, **Link options**, **Preview**, **Cell contents**, and **Cell formatting**. The options in the **File** area are used to specify the location of excel file and its path type. The options in the **Link Options** area are used to specify the data that is to be linked from the excel sheet. The **Preview** area is used to display preview of the selected data that is to be linked. The options in the **Cell contents** area are used to specify the way the content is being linked to the cell. There are three radio buttons available in this area: **Keep data formats and formulas**, **Keep data formats and solved formulas in Excel**, and **Convert data formats to text, solved formulas in Excel**. The options in the **Cell formatting** area are used to modify formatting of the current cell.

After specifying the desired settings in the dialog box, choose the **OK** button from the dialog box to apply the settings; the **Select a Data Link** dialog box will be displayed again with the newly created data link selected. Choose the **OK** button from the dialog box; the selected link will be applied to the selected cell.

### Download from Source

This tool is available in the **Data** panel of the **Table Cell** contextual tab. This tool is used to update the link from the source file. This tool is active only when you select a link from the table.

# TUTORIAL

## Tutorial 1

In this tutorial, you will create an air terminal devices schedule for the mechanical system created in exercise 1 of chapter 4, refer to Figure 8-48.

**(Expected time: 30 min)**

*Figure 8-48*  *The mechanical system created in Chapter 4*

### Opening the Drawing File in HVAC Workspace

1. Start AutoCAD MEP and choose the **Open** tool from the **Quick Access Toolbar**; the **Open** dialog box is displayed.

2. Select the drawing file created in Exercise 1 of Chapter 4 and then choose **Open** button; the file will open in the application window.

3. Click on the **Workspace Switching** button in the **Application Status Bar**; a flyout is displayed.

4. Choose the **HVAC** option from the flyout; the **HVAC** workspace is activated.

## Adding Air Terminal Devices Schedule

1.  Choose the **Air Terminal Devices Schedule** tool from the **Schedule** drop-down in the **Annotation** panel of the **Home** tab in **Ribbon**; you are prompted to select the objects to be included in the schedule.

2.  Select all the ducts available in the drawing area and then press ENTER; the schedule gets attached to the cursor.

3.  Click to specify the upper left corner of the schedule; you will be prompted to specify the lower left corner.

4.  Click to specify the lower right corner of the schedule; the schedule is placed at the specified position.

    By default, "?" marks are displayed in each of the field. You need to update the property sets in these fields.

## Updating the Schedule

1.  Select the schedule from the drawing area; the **Schedule Table** contextual tab is displayed in the **Ribbon**.

2.  Choose the **Add All Property Sets** tool from the **Modify** panel in the contextual tab; the properties in the schedule change accordingly, refer to Figure 8-49.

| | | | MECHANICAL AIR TERMINAL DEVICES SCHEDULE | | | | | |
|-----|------|-------------|----------|--------|-------------------|--------------|-----------------|-------|
| TAG | SIZE | DESCRIPTION | CONSTRUCTION | | | BASIS OF DESIGN | | NOTES |
| | | | MATERIAL | FINISH | DISCHARGE PATTERN | MANUFACTURER | MODEL OR SERIES | |
| D-1 | | | Aluminum | | 4-Way | | | |
| D-2 | | | Aluminum | | 4-Way | | | |
| D-3 | | | Aluminum | | 4-Way | | | |
| D-4 | | | Aluminum | | 4-Way | | | |
| D-5 | | | Aluminum | | 4-Way | | | |
| D-6 | | | Aluminum | | 4-Way | | | |
| D-7 | | | Aluminum | | 4-Way | | | |
| D-8 | | | Aluminum | | 4-Way | | | |

*Figure 8-49  The mechanical air terminal devices schedule*

You can also edit the schedule to include more information in the blank fields.

## Saving the Drawing

1.  Choose **Save** from the **Application** menu to save the drawing file.

## Self-Evaluation Test

Answer the following questions and then compare them to those given at the end of this chapter:

1. Vertical section is a section created by a plane perpendicular to the object to be sectioned. (T/F)

2. The **Generate Section** tool is used to create live sections by using the current section line. (T/F)

3. To create a horizontal section, choose the _____ tool from the **Section & Elevation** panel of the **Home** tab in the **Ribbon**.

4. The _____ tool is used to extract a polyline outline of the slice created through the model.

5. In which of the following panels is the **Detail Components** tool available ?

   (a) **Build**                    (b) **Details**
   (c) **Modify**                   (d) **Draw**

## Review Questions

Answer the following questions:

1. The **Detail Components** tool is not available in the **Schematic** workspace. (T/F)

2. The **Space Engineering Schedule** tool is available in the **Schedules** drop-down in the **Annotation** panel of the **Home** tab in the **Ribbon** when the **HVAC** option is selected in the **Workspace Switching** flyout. (T/F)

3. The **Space Inventory Schedule** option is available in the **Schedules** drop-down in the **Annotation** panel of the **Home** tab in the **Ribbon** when the _____ option is selected in the **Workspace Switching** flyout.

4. The _____ tool is used to refresh all the 2D sections and elevations created in the current project or in the specified folder.

5. In which of the following dialog boxes, the **Recipe** option is available?

   (a) **Generate Section/Elevation**        (b) **Batch Refresh Section/Elevations**
   (c) **Component Properties**              (d) **Schedule Table Style Properties**

# Exercise

## Exercise 1

In this exercise, you will create the 1-Phase Branch Panel schedule of the panel in the drawing, as shown in Figure 8-50. The drawing file is created in Tutorial 1 of Chapter 7. The schedule is given in Figure 8-51 for your reference.

**(Expected time: 30 min)**

*Figure 8-50* *The mechanical air terminal devices schedule*

| PANEL | Main Panel | | | | | | | | | | | |
|---|---|---|---|---|---|---|---|---|---|---|---|---|
| VOLTAGE (L–N): | | 230 | | | | ENCLOSURE TYPE: | – – – – | | | | | |
| VOLTAGE (L–L): | | 240 | | | | MOUNTING: | SURFACE | | | | | |
| PHASES, WIRES: | | 1 φ, 3 W | | | | AIC RATING: | 800 | | | | | |
| MINIMUM BUS CAPACITY (A): | | 800 A | | | | NOTES: | for light only | | | | | |
| MAIN O.C. DEVICE (A): | | 15 A | | | | | | | | | | |
| CKT NO | DESCRIPTION | TRIP AMPS | POLE | PHASE LOADS (VA) | | POLE | TRIIP AMPS | DESCRIPTION | | CKT NO | | |
| | | | | A | B | | | | | | | |
| Fans | SPARE | 20 | 1 | 720 | 750 | 1 | 20 | SPARE | | Lights | | |
| Sockets | SPARE | 20 | 1 | | 3000 | | | | | | | |
| | | | | CONNECTED LOAD PHASE TOTALS (VA) | | | | | | | | |
| | | | | 1470 | 3000 | | | | | | | |

| | CONNECTED LOAD (KVA) | DEMAND FACTOR | DEMAND LOAD (KVA) | DEMAND LOAD | 4.7 KVA |
|---|---|---|---|---|---|
| Motors | 0.5 | 1.00 | 0.5 | SPARE CAPACITY | 0.1KVA |
| Motors (Largest) | 0.2 | 1.25 | 0.2 | SPARE CAPACITY | 0.4 AMPS |
| Sockets (0 – 10 KVA) | 3.0 | 1.00 | 3.0 | SPARE CAPACITY | 2 % |
| Lighting | 0.8 | 1.25 | 0.9 | | |
| TOTAL: | 4.5 | | 4.7 | | |
| LOAD (AMPS): | 18.6 | | 19.6 | | |

*Figure 8-51* *The 1-Phase Branch Panel schedule for the drawing*

## Answers to Self-Evaluation Test
1. T, 2. F, 3. Horizontal Section, 4. Quick Slice, 5. Details

# Chapter 9

# Working with Schematics

# INTRODUCTION

In this chapter, you will learn about various tools that are used for creating a schematic drawing for a building system. You will also learn the use of equipment and schematic lines in a schematic drawing. The tools to create a schematic drawing are available in the **Schematic** workspace. This workspace is discussed next.

# SCHEMATIC WORKSPACE

To invoke this workspace, choose the **Schematic** option from the **Workspace Switching** flyout; the **Schematic** workspace will be activated. Figure 9-1 shows a partial view of the **Ribbon** in the **Schematic** workspace. The tools to create a schematic drawing are available in the **Build** panel of the **Home** tab in the **Ribbon**. These tools are discussed next.

*Figure 9-1  Partial view of the **Ribbon** in the **Schematic** workspace*

## Equipment

Equipment  This tool is available in the **Build** panel of the **Home** tab in the **Ribbon**. This tool is used to add equipment in the drawing. To add an equipment, choose the **Equipment** tool from the **Ribbon**; the **Add Multi-view Parts** dialog box will be displayed. The options in this dialog box are the same as discussed in Chapter 6.

## Schematic Symbol

Schematic Symbol  Schematic symbols are the symbolic representations of physical objects required in a building system. In AutoCAD MEP, there are two types of schematic symbols: In-line symbols and End-of-line symbols. The In-line symbols are those that can be added in-between the schematic line. Note that when you delete a schematic line in which the In-line symbols are attached, the symbols also get deleted with the line. The End-of-line symbol are those that can be added either at the start point or at the end point of a schematic line. To add a schematic symbol, choose the **Schematic Symbol** tool from the **Build** panel of the **Home** tab in the **Ribbon**; the symbol will get attached to the cursor and you will be prompted to specify the insertion point for the symbol. Also, the **Properties** palette will be displayed, as shown in Figure 9-2. Click in the drawing area to place the symbol. The options in the **Properties** palette are discussed next.

*Figure 9-2  The **Properties** palette displayed on choosing the **Schematic Symbol** tool*

## Properties Palette

The options in the **Properties** palette are used to change the parameters of the schematic symbols. These options are discussed next.

### Description

This field is available in the **BASIC > General** rollout of the **Design** tab in the **Properties** palette. When you click on this field, the **Description** dialog box will be displayed, as shown in Figure 9-3. Enter description for the symbol in the **Edit the description for this object** text box in this dialog box. Choose **OK** to exit the dialog box.

*Figure 9-3  The **Description** dialog box*

### Style

This option is available in the **BASIC > General** rollout of the **Properties** palette. This option is used to select a schematic symbol from the list of symbols available in AutoCAD MEP. To select a symbol, choose the **Style** option from the **Properties** palette; the **Select a style** dialog box will be displayed, as shown in Figure 9-4. In this dialog box, the options in the **Drawing file** drop-down list are used to specify the template file for schematic symbols. If you select any option from the **Drawing file** drop-down list, the categories available in the respective template file will be displayed in the **Category** drop-down list. Select the desired category from the **Category** drop-down list; the respective symbols will be displayed in the dialog box. Select a symbol from the dialog box and then choose the **OK** button to exit.

*Figure 9-4  The **Select a style** dialog box*

### System
This drop-down list is available in the **BASIC > System** rollout. This drop-down list contains various system definitions for building systems. The option selected in this drop-down list specifies the system to which the symbol belongs.

### Orientation
This drop-down list is available in the **BASIC > Drawing Mode** rollout. The options in this rollout are used to specify the orientation in which the symbol will be placed in the drawing. There are two options available in this drop-down list: **Ortho** and **Isometric**. The **Ortho** option is used to place the symbols in orthographic mode, while the **Isometric** option is used to place the symbol in isometric orientation. The **Isometric** option is active only when an isometric schematic symbol is selected in the **Select a style** dialog box.

### Isoplane
This drop-down list is available in the **Properties** palette only when the **Isometric** option is selected in the **Orientation** drop-down list. There are three options in this drop-down list: **Left**, **Right**, and **Top**. Select an option from this drop-down list; the symbol will be oriented along the plane corresponding to the selected option.

### Justification

This drop-down list is available in the **Location** rollout. The options in this drop-down list are used to position the symbol with respect to the insertion point.

### Rotation

This edit box is available in the **Location** rollout for the End-of-line symbols in orthographic mode. To rotate the symbol, enter the desired value of rotation in this edit box.

### Rotation in isoplane

This field is available in the **Location** rollout for the symbols in isometric mode. To active this field, you need to specify the insertion point for the symbol. The value in this field changes according to the position of the cursor.

### ID

This edit box is also available in the **ADVANCED > Design Data** rollout in the **Properties** palette. You can assign a unique ID to the symbol being inserted in the drawing area.

After placing a symbol in the drawing area, you can change the orientation of the symbol dynamically. Figure 9-5 shows the **2 Port Pneumatic Valve** symbol placed in the isometric mode with its orientation handles. The handles displayed on the symbols are discussed next.

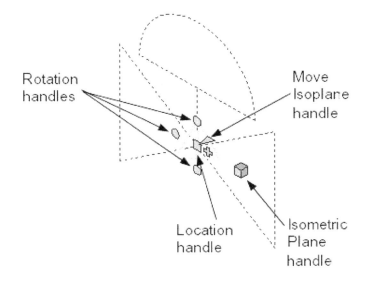

*Figure 9-5* *The **2 Port Pneumatic Valve** symbol with its orientation handles*

### Rotation handles

These handles are used to rotate the symbol. When you click on a handle, the symbol is rotated with respect to the insertion point.

**Location handle**
This handle is used to modify the location of the symbol. When you click on this handle and move the cursor, the symbol moves along the cursor in the selected isoplane.

**Move Isoplane handle**
This handle is used to change the location of the symbol perpendicular to the isoplane selected.

**Isometric Plane handle**
This handle is used to change the current isometric plane of the symbol. Click on this symbol to switch between Left, Right, and Top planes.

# Schematic Line

Schematic lines are used to represent various types of connections in a building system. To create a schematic line, choose the **Schematic Line** tool from the **Build** panel in the **Home** tab of the **Ribbon**; you will be prompted to specify the start point for the schematic line. Also, the **Properties** palette will be displayed, as shown in Figure 9-6. Click to specify the start point

*Figure 9-6 The **Properties** palette displayed on choosing the **Schematic Line** tool*

of the line; you will be prompted to specify the next point of the schematic line. Click to specify the second point of the line; you will be again prompted to specify the next point of the line. Click to specify the third point or press ENTER to exit the tool. Except the **Length** edit box, the options displayed in the **Properties** palette on choosing the **Schematic Line** tool have already been discussed.

The **Length** edit box is used to specify the length of the schematic line. You cannot modify the value of length once you have created the schematic line. Also, you cannot specify the value of length directly in this edit box. To specify a value in this edit box, you need to specify the desired value in the dynamic prompt displayed along the cursor while creating the schematic line.

## Schematic Line Styles

While creating a schematic diagram, you might require to change the style of the schematic lines. The style of schematic lines can be changed by using the **Style Manager**. To invoke the **Style Manager** for modifying the schematic line style, choose the **Schematic Line Styles** tool from the **Style Manager** drop-down in the **Style & Display** panel of the **Manage** tab in the **Ribbon**; the **Style Manager** will be displayed, as shown in Figure 9-7. Select a line style from the list available in the left of the dialog box; the options related to that line style will be displayed in the right of the dialog box, refer to Figure 9-8. The options in the **Style Manager** after selecting a line style are discussed next.

*Figure 9-7* The **Style Manager** with schematic line styles

*Figure 9-8* *The **Style Manager** with options related to a line style*

## General Tab

This tab is selected by default in the **Style Manager**, refer to Figure 9-8. The options in this tab are used to change the general settings for the line style. These options are discussed next.

### Name

This edit box is used to change the name of the schematic line style selected.

### Description

This edit box is used to specify the description about the selected line style.

### Keynote

This edit box is used to specify key notes for the schematic line style. You can enter the desired key notes in the edit box or you can choose the **Select Keynote** button next to the edit box to select a keynote from the list of predefined keynotes. On choosing this button, the **Select Keynote** dialog box will be displayed, refer to Figure 9-9. To specify any predefined keynote, open the desired category by clicking on the plus sign adjacent to the category and double-click on the desired keynote; the name of the selected keynote will be displayed in the **Keynote** edit box.

*Figure 9-9  The Select Keynote dialog box*

The description of the keynote is also displayed in the edit box available below the **Keynote** edit box.

### Notes

This button is used to specify notes for the selected line style. To specify the notes, choose the **Notes** button from the bottom of the **General** tab; the **Notes** dialog box will be displayed, as shown in Figure 9-10. There are two tabs available in this dialog box: **Notes** and **Reference Docs**. On choosing the **Notes** tab, you can specify notes regarding the line style in the edit box displayed below it. You can also add a reference document to the selected line style. To do so, choose the **Reference Docs** tab; the **Notes** dialog box will be displayed, as shown in Figure 9-11. Choose the **Add** button from the dialog box; the **Select Reference Document** dialog box will be displayed, as shown in Figure 9-12.

*Figure 9-10*  *The Notes dialog box*

*Figure 9-11*  *The Notes dialog box with the Reference Docs tab chosen*

*Figure 9-12  The **Select Reference Document** dialog box*

Browse to the desired location, select the document to be attached, and choose the **OK** button from the dialog box; the **Reference Document** dialog box will be displayed, as shown in Figure 9-13. Enter the description about the document to be added in the **Description** edit box and then choose the **OK** button from the dialog box; the selected file will be displayed in the list of documents in the **Notes** dialog box. Choose the **OK** button from the dialog box; the selected file will be added as note for the selected line style. You can add more documents by choosing the **Add** button from the **Notes** dialog box. You can edit by choosing the **Edit** button. To delete the document, choose the **Delete** button.

*Figure 9-13  The **Reference Document** dialog box*

### Property Sets

This option is used to specify property set data for the selected style. Property sets are the parameters that can be changed for each instance of the object created using the specific style.

## Designations Tab

The options in this tab are used to specify designations for the schematic lines, refer to Figure 9-14. Designations are used to create unique IDs for the schematic lines. To specify designations, choose the **New Designation (ID)** tool 📧 from the tab; **1** will be displayed under the **Index** column of the list. Click in the field corresponding to **1** in the **Designation (ID)** column and specify the desired designation in the field. You can also create more than one designation for a schematic line by choosing the **New Designation (ID)** tool.

*Figure 9-14  The Style Manager with the Designations tab chosen*

## Annotation Tab

The options in this tab are used to modify annotation style for a schematic line, refer to Figure 9-15. The options available in this tab are discussed next.

### Crossings Area

There are three buttons available in this area: **Do Nothing**, **Overlap Graphics**, and **Break Existing Line**. The **Do Nothing** button is used to create the crossing at the intersection point of the schematic lines. The **Overlap Graphics** button is used to create an overlapping graphic at the intersection point of the schematic lines. The **Break Existing Line** button is used to break the existing line at the intersection point. The **A - Break/Overlap Paper Width** edit box, available adjacent to the **Break Existing Line** button, is used to change the width of overlap or break. The **Break/Overlap Priority** edit box is available below the **A - Break/Overlap Paper Width** edit box. This edit box is used to set overlap or break priority.

*Figure 9-15  The Style Manager with the Annotations tab chosen*

### Connections Area
The **Connection Node** drop-down list in this area is used to specify the style of nodes created at the connections. You can specify the size of connection node in the **Paper Size** edit box.

### Start & End Settings Area
There are two drop-down lists in this area: **Start** and **End**. The options in this drop-down lists are used to specify the shape nodes created at the start and end of the schematic lines. You can specify the sizes of the nodes in the corresponding **Paper Size** edit boxes.

## Display Properties Tab
The options in this tab are used to modify the display representations of schematic lines, refer to Figure 9-16. There are various templates for display representations available in this tab. You can override any of the representations by using the options available in this tab.

*Figure 9-16  The **Style Manager** with the **Display Properties** tab chosen*

## Version History Tab

The options in this tab are used to record the history of the selected line style, refer to Figure 9-17. You can check the versions of the selected line style with the date of modification listed at the bottom of the dialog box. You can also remove any of the versions by using the **Remove** button.

*Figure 9-17  The **Style Manager** with the **Version History** tab chosen*

There are some common tools available in the **Style Manager** to create or edit line styles. These tools are discussed next.

### New Drawing
This tool is used to add a new drawing in the **Style Manager**. You can add various line styles in this drawing by copying them.

### Open Drawing
This tool is used to add an existing drawing in the **Style Manager**. Various styles available in the drawing are automatically imported in the **Style Manager**.

### Copy
This tool is used to copy the selected line style. To do so, select a line style from the list at the left in the **Style Manager** and then choose the **Copy** tool from the **Style Manager**; the style will be copied in the clipboard.

### Paste
This tool is used to paste the line style copied from the list by using the **Copy** tool. To paste a line style, choose the **Paste** tool from the **Style Manager**; the line style will be pasted in the selected drawing.

### Edit Style
This tool is used to switch to the editing mode.

### New Style
This tool is used to create a new line style. To do so, choose the **New Style** tool from the toolbar displayed in the **Style Manager**; a new line style will be created and added in the list with the name **New Style**. Also, you will be prompted to change the name of the style. The options corresponding to the style will be displayed at the right in the **Style Manager**.

### Set From
This tool is used to set the shape of the style by using an existing drawing file.

### Purge Styles
This tool is used to remove the unused styles from the list displayed in the **Style Manager**. To remove styles, select the styles to be removed and then choose the **Purge Styles** tool from the toolbar; the selected styles will be removed from the list.

## Toggle View

 This tool is used to toggle between the display styles available in the **Style Manager**. There are two display styles available in the **Style Manager**: Display per category and Display per drawing. The Display per drawing display style is selected by default in the **Style Manager**. If this style is selected, the line styles are displayed as per the drawings. If the Display per category style is selected, the styles will be displayed according to their categories.

## Filter Style Type

 This tool is used to toggle between the display styles available in the **Style Manager**. If this tool is chosen, then the styles corresponding only to the selected type will be displayed.

## Inline Edit Toggle

This tool is used to toggle between the symbolic view/list and properties of a line style. On choosing this toggle button, the options in the **Style Manager** will get modified, as shown in Figure 9-18.

There are two tabs in the right area of the dialog box, the **Viewer** tab and the **List** tab. The options in the **Viewer** tab are used to view the current line style in various orientations. On choosing the **List** tab, the options in the **Style Manager** will get modified, refer to Figure 9-19. Also, the information regarding the selected line style will be displayed in this tab.

*Figure 9-18  The Style Manager after choosing the Inline Edit Toggle button with the Viewer tab chosen*

*Figure 9-19 The **Style Manager** after choosing the **Inline Edit Toggle** button with the **List** tab chosen*

## Schematic Representation of an Existing System

MEP Design ▾ You can display the schematic representation of any system created in other workspaces. To display the schematic representation, open the desired file and then click on the **Level of Details** button in the **Drawing Status Bar**; a flyout will be displayed, as shown in Figure 9-20.

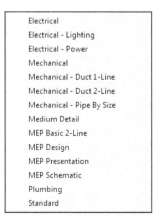

*Figure 9-20 The flyout displayed on choosing the **Level of Details** button*

Choose the **MEP Schematic** option from the flyout; the current drawing will be displayed with schematic representations, refer to Figure 9-21.

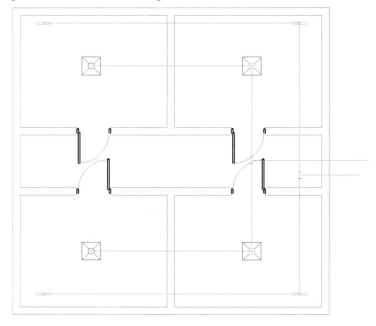

*Figure 9-21* *The schematic representation of an HVAC system*

# TUTORIAL

## Tutorial 1

In this tutorial, you will create a schematic diagram of a computer lab, refer to Figure 9-22.

**(Expected time: 30 min)**

*Figure 9-22* *The schematic representation of a computer lab*

Examine the model to determine schematic symbols to be added.

The following steps are required to complete this tutorial:

a. Start a new drawing in **Schematic** workspace.
b. Add the schematic symbols.
c. Change the system of the symbols.
d. Connect the symbols with schematic lines.

## Starting a New Drawing File in Schematic Workspace

1. Start AutoCAD MEP, a new drawing file with the default name *Drawing1* is opened.

2. Click on the **Workspace Switching** button in the **Application Status Bar**; a flyout is displayed.

3. Choose the **Schematic** option from the flyout; the **Schematic** workspace is activated.

## Adding Schematic Symbols

1. Choose the **Schematic Symbol** tool from the **Build** panel of the **Home** tab in the **Ribbon**; the **Properties** palette is displayed, as shown in Figure 9-23. Also, you are prompted to specify the location of the symbol.

*Figure 9-23 The **Properties** palette after choosing the **Schematic Symbol** tool*

2. Click in the **Style** field of the **Properties** palette; the **Select a style** dialog box is displayed, as shown in Figure 9-24.

*Figure 9-24* The **Select a style** dialog box

3. Select the **Electrical Communications (Global)** option from the **Drawing file** drop-down list and the **Telephone and Data** option from the **Category** drop-down list; the **Select a style** dialog box is modified, as shown in Figure 9-25.

*Figure 9-25* The modified **Select a style** dialog box

4.  Select the **Telephone Point** symbol from the list displayed in the dialog box and choose the **OK** button; the symbol gets attached to the cursor.

5.  Click in the drawing area to place the symbol; you are prompted to specify the rotation value.

6.  Specify the value of rotation as **0** at the command prompt; the symbol is placed at the specified location. Press ESC to exit the tool.

7.  Select the symbol from the drawing area and right-click; a shortcut menu is displayed. Choose the **Array** tool from the **Basic Modify Tools** flyout.

8.  Choose the **Rectangular** option from the command prompt; the preview of the array of telephone points is displayed, refer to Figure 9-26. Also, the **Array Creation** contextual tab is displayed, as shown in Figure 9-27.

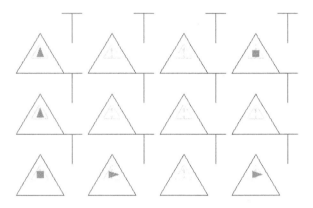

*Figure 9-26* *The preview of the array of telephone points*

*Figure 9-27* *The Array Creation contextual tab*

9.  Specify the value **4** in the **Columns** edit box and **3000** in the **Between** edit box in the **Columns** panel of the contextual tab; the preview of the array is displayed.

10. Specify the value **2** in the **Rows** edit box and **3000** in the **Between** edit box in the **Rows** panel; the preview of the array is displayed, as shown in Figure 9-28.

*Figure 9-28  The preview of array*

11. Choose the **Close Array** tool from the **Close** panel of the contextual tab; the array of telephone points is created.

12. Similarly, add other symbols according to Figure 9-22. The drawing after adding other symbols is displayed, refer to Figure 9-29.

 **Note**

1. *Symbol for **PC** is available in the **Equipment** category of the **Electrical Power (Global)** drawing file in the **Select a style** dialog box.*

2. *Symbol for **240v-110v Switched Socket Outlet** is available in the **Sockets** category of the **Electrical Power (Global)** drawing file in the **Select a style** dialog box.*

3. *Symbol for **Inline Fuse Electrical** is available in the **Switches** category of the **Electrical Schematic (Global)** drawing in the **Select a style** dialog box.*

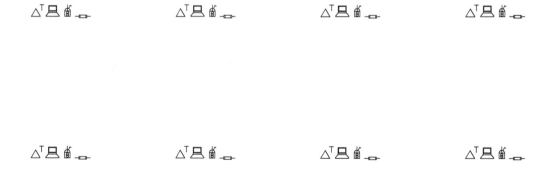

*Figure 9-29  The drawing after adding other symbols*

## Changing the System of Symbols

Now, you need to change the system of symbols available in the drawing area as per their application.

1. Select all the symbols of **PC**, **240v-110v Switched Socket Outlet**, and **Inline Fuse Electrical** from the drawing; the modified **Properties** palette is displayed, as shown in Figure 9-30.

*Figure 9-30  The modified* **Properties** *palette*

2. Select the **E-230V Power (230-PWR)** option from the **System** drop-down list in the **System** rollout of the **Properties** palette and select **Z-Schematics-G** from the **Layer** drop-down list in the **General** rollout. The **Properties** palette after specifying the parameters is displayed, as shown in Figure 9-31.

3. Similarly, select all the Telephone Points in the **Properties** palette, and then select the **E-Telephone (TELEPHONE)** option from the **System** drop-down list. Also, select the **Z-Schematics-G** option from the **Layer** drop-down list.

## Connecting Symbols with Schematic Line
Now, you need to connect all the schematic symbols with their relevant schematic lines.

1. Choose the **Schematic Line** tool from the **Build** panel in the **Home** tab of the **Ribbon**; the **Properties** palette is displayed, as shown in Figure 9-32.

*Figure 9-31* The **Properties** *palette after specifying the parameters*

*Figure 9-32* The **Properties** *palette after choosing the* **Schematic Line** *tool*

2.  Select the **E-230V Power (230-PWR)** option from the **System** drop-down list and the **Wire** option from the **Style** drop-down list.

3.  Click on **Schematic End Connector** of the **PC** at the top-left in the drawing, refer to Figure 9-33; the schematic line gets attached to the cursor.

*Figure 9-33* *The schematic end connector to be selected*

4. Move the cursor towards right and select the Schematic End Connector of the 240v-110v Switched Socket Outlet adjacent to the selected PC. Similarly, connect the Inline Fuse Electrical in the drawing area.

5. Similarly, connect rest of the symbols, refer to 9-34.

*Figure 9-34* *The drawing after connecting all the symbols of the* **Electrical Power (Global)** *and* **Electrical Schematic (Global)** *categories*

6. Choose the **Schematic Line** tool again and select the **E-Telephone (TELEPHONE)** option from **System** drop-down list in the **Properties** palette; you will be prompted to specify the start point of the line.

7. Click on the Schematic End Connector of the Telephone point; the schematic line gets attached to the cursor.

8. Connect all the telephone points with the schematic lines, as shown in Figure 9-35.

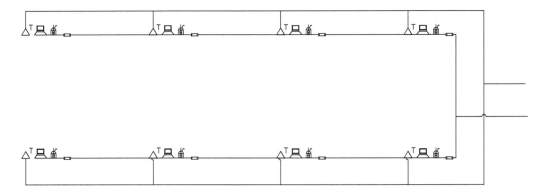

*Figure 9-35* *The drawing after connecting all the telephone points with the schematic lines*

## Self-Evaluation Test

**Answer the following questions and then compare them to those given at the end of this chapter:**

1.  Schematic symbols are the symbolic representation of physical objects required in a building system. (T/F)

2.  When you delete a schematic line in which the In-line symbols are attached, the symbols do not get deleted with the line. (T/F)

3.  The **System** drop-down list is available in the _____ rollout of the **Design** tab in the **Properties** palette.

4.  The _____ handle is used to modify the location of the symbol.

5.  In which of the following drawing files, the **360 Camera** symbol is available in the **Select a style** dialog box ?

    (a) **Electrical Communications (Global)**      (b) **Electrical Lighting (Global)**
    (c) **Mechanical Equipment (Global)**           (d) **Electrical Schematic (Global)**

## Review Questions

**Answer the following questions:**

1. The **Ortho** option is used to place the symbols perpendicular to the insertion point. (T/F)

2. The style of schematic lines can be changed by using the **Style Manager**. (T/F)

3. The _____ option is used to select a schematic symbol from the list of symbols available in AutoCAD MEP.

4. The _____ option in the **Level of Details** flyout is used to display the schematic representation of a drawing.

5. In which of the following drawing files, the **Card Reader** symbol is available in the **Style** dialog box ?

    (a) **Electrical Communications (Global)**    (b) **Electrical Lighting (Global)**
    (c) **Mechanical Equipment (Global)**    (d) **Electrical Schematic (Global)**

# Exercise

## Exercise 1

In this exercise, you will create a schematic drawing, as shown in Figure 9-36. The schematic line for connecting the symbols should be of the **E-230V Power (230-PWR)** system.

**(Expected time: 30 min)**

*Figure 9-36  Schematic drawing for Exercise 1*

# Chapter 10

# Project

**Learning Objectives**

*After completing this chapter, you will be able to:*
- *Create a project file*
- *Add architectural drawing in the project*
- *Add mechanical system in the project*
- *Add electrical system in the project*
- *Add piping system in the project*
- *Add plumbing system in the project*

# PROJECT

In this project, you will create a complete system of a forging plant. Figure 10-1 shows the architectural drawings of the plant. Figure 10-2 shows the HVAC drawings of the plant. Figure 10-3 shows the piping drawings of the plant. Figure 10-4 shows the electrical drawings of the plant. Table 10-1 shows the parameters required for various devices.

**(Expected time: 30 min)**

***Figure 10-1*** *Architectural drawings of the plant*

Ground Floor

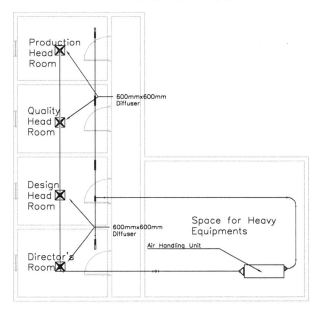

First Floor

*Figure 10-2* *HVAC drawings of the plant*

Ground Floor

First Floor

*Figure 10-3* Piping drawings of the plant

Ground Floor

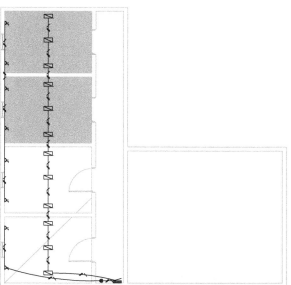

First Floor

**Figure 10-4** *Electrical drawings of the plant*

*Table 10-1* *The parameters required for various devices*

| Device | 300x1200 Surface Type A Light | 300x600 Recessed Light | Single Switch Socket |
|---|---|---|---|
| Load | 180VA | 100 | 300VA |
| Voltage | 230 | 230 | 230 |
| Phases | 1 | 1 | 1 |
| Maximum Overcurrent Rating | 10 | 8 | 10 |
| Power Factor | 0.8 | 0.8 | 0.8 |

The following steps are required to complete this project:

1. Start a new project file
2. Create categories in the project
3. Create levels in the project
4. Create architectural drawings
   - Create architectural drawing of ground floor
   - Create walls in the architectural drawing for ground Floor
   - Change the door style in the architectural drawing
   - Create doors in the architectural drawing of ground floor
   - Create windows in the architectural drawing of ground floor
   - Create architectural drawing of first floor
   - Create walls in the architectural drawing for first floor
   - Create doors in the architectural drawing of first floor
   - Create windows in the architectural drawing of first floor
5. Create HVAC drawings
   - Create HVAC drawing of ground floor
   - Import the architectural drawing of ground floor
   - Add air terminals in the architectural drawing of ground floor
   - Create the duct line of ground floor
   - Calculate the duct sizes of ground floor
   - Create return duct for the ground floor
   - Create return duct line for the ground floor
   - Calculate the duct sizes of ground floor
   - Create HVAC drawing of first floor
   - Import the architectural drawing of first floor
   - Add air terminals in the architectural drawing of first floor
   - Add duct line to the diffusers
   - Add air handling unit
   - Add return grilles in the architectural drawing of first floor
   - Create return duct line for the first floor
   - Calculate the duct sizes of first floor
6. Create piping drawings
   - Create piping drawing of ground floor
   - Import the architectural drawing of ground floor
   - Add ball valves in the architectural drawing of ground floor
   - Add pipe line in the architectural drawing of ground floor

- Connect ball valves with the pipe line
- Create piping drawing of first floor
- Import the architectural drawing of first floor
- Add tank in the first floor
- Add pumps in the first floor
- Add pipe line in the first floor

7. Create electrical drawings
- Create electrical drawing of ground floor
- Import the architectural drawing of ground floor
- Add **300x1200 surface type lights** in the drawing
- Add sockets in the drawing
- Add **300x600 recessed lights** in the drawing
- Configure devices of ground floor
- Create panel of ground floor
- Configure circuits of ground floor
- Add wires of ground floor
- Calculate load and wire sizes of ground floor
- Add transformer and emergency generator to the drawing based on total load
- Create electrical drawing of first floor
- Import the architectural drawing of first floor
- Add lights in the drawing of first floor
- Add sockets in the drawing of first floor
- Configure devices of first floor
- Create panel of first floor
- Configure circuits of first floor
- Add wires of first floor
- Calculate loads and wire sizes of first floor

## Starting a New Project File

1. Start AutoCAD MEP by using the *AutoCAD MEP 2014 - English (Global)* icon displayed on your desktop.

2. Choose **New > Project** from the **Application** menu; the **Project Browser** is displayed.

3. Choose the **New Project** button available at the bottom of the **Project Browser**; the **Add Project** dialog box is displayed.

4. Enter the project number, project name, and description in their respective fields. Enter the project name as **Forging Plant** in the **Project name** field, 0001 in the Project Number field. Make sure that the **Create from template project** check box is clear.

5. Choose the **OK** button; the newly created project file name is displayed in the left pane of the **Project Browser**, refer to Figure 10-5.

6. Choose the **Close** button from the **Project Browser**; the **Project Navigator** palette is displayed in the drawing area and the new project gets activated.

# Creating Categories in the Project

Now, you need to categorize the project.

1.  To categorize a project, choose the **Add Category** tool from the bottom of the **Project Navigator**; a new category is added in the **Constructs** node.

*Figure 10-5  The **Project Browser***

2.  Specify the name of the category as **Architectural** and press ENTER.

3.  Similarly, add other categories and name them as **Mechanical**, **Piping**, and **Electrical**.

# Creating Levels in the Project

Now, you need to create levels in the project.

1.  To create levels, choose the **Project** tab from the **Project Navigator**, the **Project Navigator** is displayed, as shown in Figure 10-6.

2.  Choose the **Edit Levels** button from the **Levels** rollout of the **Project Navigator**; the **Levels** dialog box is displayed, as shown in Figure 10-7.

3.  Enter **4000** in the **Floor to Floor Height** column respective to **0** in the **Name** column.

4.  Choose the **Add Level** button from the right of the dialog box; a new level is added with the name **1** in the list displayed in the dialog box.

5.  Choose the **OK** button from the dialog box to exit. If you get a message box, as shown in Figure 10-8, then choose the **Yes** button from the message box.

*Figure 10-6*  The **Project Navigator** *with the* **Project** *tab chosen*

*Figure 10-7*  The **Levels** *dialog box*

*Figure 10-8*  The message box

# Creating Architectural Drawings

Now, you need to create architectural drawings for the ground floor and first floor.

## Creating Architectural Drawing of the Ground Floor

1. Select the **Architectural** category from the **Constructs** tab in the **Project Navigator** and then choose the **Add Construct** button available at the bottom; the **Add Construct** dialog box is displayed, as shown in Figure 10-9.

2. Select the **Open in drawing editor** check box from the bottom of the dialog box.

*Figure 10-9 The **Add Construct** dialog box*

3. Enter **Ground Floor Architectural** in the **Name** field of the dialog box.

4. Click in the **Description** field; the **Description** dialog box is displayed, as shown in Figure 10-10.

5. Enter the description as **This drawing is meant for architectural plan of the ground floor** in the **Edit the description** edit box and then choose the **OK** button from the dialog box.

6. Select the check box corresponding to level **0** in the **Division** column in the **Add Construct** dialog box and then choose the **OK** button from the dialog box; the drawing is opened

**Project**

**10-11**

in AutoCAD MEP. Also, the Ground Floor Architectural drawing is displayed with a lock icon adjacent to it in the **Project Navigator**.

*Figure 10-10  The **Description** dialog box*

7.  Click on the **Workspace Switching** option in the Drawing Status Bar and select the **Architecture** option from the flyout displayed; the **Architecture** workspace is activated.

## Creating Walls in the Architectural Drawing

1.  Choose the **Wall** tool from the **Wall** drop-down in the **Build** panel of the **Home** tab in the **Ribbon**; you are prompted to specify the starting point of the wall. Also, the **Properties** palette is displayed, refer to Figure 10-11.

*Figure 10-11  The **Properties** palette*

2.  Enter the base height of the wall as **4000**, width as 254, and set the **Justify** option as **Right** in the **Properties** palette.

3. Enter the coordinates **0,0,0** at the command prompt to specify the start point; a rubber band wall is displayed with its other end attached to the cursor.

4. Press F8 to work in the Ortho Mode and specify the horizontal length of the wall **18000** at the command prompt.

5. Move the cursor vertically upward and specify the value **18000** at the command prompt.

6. Move the cursor towards the left and specify the value **18000** at the command prompt.

7. Enter **C** at the command prompt to create the last wall. The drawing displays the architectural outer boundary of the plant, refer to Figure 10-12.

*Figure 10-12   The architectural outer boundary of the plant*

8. Similarly, create the walls inside the plant, refer to Figure 10-13.

## Changing the Door Style in the Architectural Drawing

1. Choose the **Style Manager** tool from the **Style & Display** panel of the **Manage** tab in the **Ribbon**; the **Style Manager** is displayed, as shown in Figure 10-14.

*Figure 10-13* *The drawing after creating all walls*

*Figure 10-14* *The Style Manager*

2.  Click on the plus sign adjacent to **Architectural Objects** in the left of the **Style Manager**; a list of architectural objects is displayed under the **Architectural Objects** category.

3. Click on the plus sign adjacent to **Door Style** in the list; the **Standard** door style is displayed in the list.

4. Select the **Standard** door style; the options related to the door style are displayed on the right in the **Style Manager**, refer to Figure 10-15.

*Figure 10-15*   *The* ***Style Manager*** *with door style selected*

5. Choose the **Design Rules** tab in the **Style Manager**; the options related to the shape of the door are displayed.

6. Select the **Rectangular** option from the **Predefined** drop-down list of the **Shape** area in the **Style Manager**.

7. Select the **Double** option from the **Door Type** area.

8. Choose the **Standard Sizes** tab; the options related to the size of door are displayed on the right in the **Style Manager**.

9. Choose the **Add New Standard Size** button from the right of the **Style Manager**; the **Add Standard Size** dialog box is displayed, as shown in Figure 10-16.

*Figure 10-16*   *The* ***Add Standard Size*** *dialog box*

10. Enter the description as **Inner Door** in the **Description** edit box of the dialog box.

11. Specify the width and height as **1000** and **3000** in the respective fields of the dialog box and then choose the **OK** button.

12. Choose the **Add New Standard Size** button again and create a door style with width and height as **4000** and **3500**, respectively. Specify **Main Door** in the **Description** edit box.

13. Choose the **OK** button to exit the **Style Manager**.

## Creating Doors in the Architectural Drawing of the Ground Floor

1. Choose the **Door** tool from the **Door** drop-down in the **Build** panel of the **Home** tab in the **Ribbon**; you are prompted to specify the insertion point of the door. Also, the **Properties** palette is displayed, as shown in Figure 10-17.

2. Select the **1000 X 3000** option from the **Standard Sizes** drop-down list in the **Dimensions** rollout of the **Properties** palette.

3. Select the **Offset** option from the **Position** drop-down list in the **Location** rollout and specify **150** in the **Automatic offset** edit box.

4. Click in the drawing area; the door gets attached to the cursor. Add all the inner doors, as shown in Figure 10-18. Press ENTER to exit the tool.

*Figure 10-17* The **Properties** palette displayed on choosing the **Door** tool

*Figure 10-18* *The drawing after adding inner doors*

5.  Again, choose the **Door** tool; you are prompted to select a wall or grid. Also, the **Properties** palette is displayed.

6.  Select the **4000 X 3500** option from the **Standard Sizes** drop-down list in the **Dimensions** rollout and click in the drawing; the door is attached to the cursor.

7.  Click at the walls to create main doors, refer to Figure 10-19.

*Figure 10-19* *The drawing after adding main doors*

## Creating Windows in the Architectural Drawing of the Ground Floor

1. Choose the **Window** tool from the **Window** drop-down in the **Build** panel of the **Home** tab in the **Ribbon**; you are prompted to select a wall or a grid assembly. Also, the **Properties** palette is displayed, as shown in Figure 10-20.

2. Choose the **Center** option from the **Position** drop-down list of the **Location** rollout in the **Properties** palette and choose the window size **1010 X 1510** from the **Standard sizes** drop-down list in the **Dimensions** rollout.

3. Click on the walls of the ground floor architectural drawing, shown in Figure 10-21, to add windows.

4. Save the drawing and close it.

*Figure 10-20  The **Properties** palette on choosing the **Window** tool*

## Creating Architectural Drawing of First Floor

1. Select the **Architectural** category from the **Project Navigator** and choose the **Add Construct** tool from the bottom; the **Add Construct** dialog box is displayed.

2. Specify the name as **First Floor Architectural** in the **Name** edit box of the dialog box.

3. Click in the **Description** field and specify the description as **This drawing is meant for architectural plan of the first floor** in the dialog box displayed. Make sure that the **Open in drawing editor** check box is selected.

4. Select the check box corresponding to level 1 in the **Division** column of the **Add Construct** dialog box and then choose the **OK** button to exit; the drawing is displayed in the drawing editor. Make sure that the **Architecture** option is selected in the **Workspace Switching** flyout.

## Creating Walls in the Architectural Drawing

1. Choose the **Wall** tool from the **Wall** drop-down in the **Build** panel of the **Home** tab in the **Ribbon**; you are prompted to specify the starting point of the wall. Also, the **Properties** palette is displayed.

2. Enter **0,0,0** at the command prompt to specify the starting point of the wall; move the cursor vertically upward and enter **18000** at the command prompt.

*Figure 10-21  The walls to be selected for adding windows*

3. Move the cursor toward right and then enter **8000** at the command prompt.

4. Move the cursor vertically downward and then enter **9000** at the command prompt.

5. Move the cursor toward right and then enter **10000** at the command prompt.

6. Move the cursor vertically downward and enter **9000** at the command prompt.

7. Enter **C** at the command prompt to create the last wall.

The outer boundary of the first floor is created, refer to Figure 10-22. Now, you need to create rooms in the building.

8.  Similarly, using the **Wall** tool, create the rooms, as shown in Figure 10-23.

*Figure 10-22  Outer boundary of the first floor*

*Figure 10-23  Drawing of first floor after creating rooms*

## Creating Doors in the Architectural Drawing of the First Floor

1. Choose the **Door** tool from the **Door** drop-down in the **Build** panel of the **Home** tab in the **Ribbon**; you are prompted to select a wall or a grid. Also, the **Properties** palette is displayed.

2. Specify width as **1500** in the **Width** edit box and height as **3000** in the **Height** edit box of the **Dimensions** rollout in the **Properties** palette.

3. Select the **Center** option from the **Position along wall** drop-down list in the **Location** rollout of the **Properties** palette.

4. Click on a wall; the door gets attached to the cursor and can move along the walls.

5. Click on the wall again; the door is added. Similarly, add rest of the doors, refer to Figure 10-24.

*Figure 10-24 Drawing of the first floor after creating doors*

## Creating Windows in the Architectural Drawing of the Ground Floor

1. Choose the **Window** tool from the **Window** drop-down in the **Build** panel of the **Home** tab in the **Ribbon**; you are prompted to select a wall or a grid assembly. Also, the **Properties** palette is displayed.

2. Select the **Center** option from the **Position along wall** drop-down list of the **Location** rollout in the **Properties** palette and then choose the window size **1010 X 1510** from the **Standard sizes** drop-down list in the **Dimensions** rollout.

3. Click on the walls, as shown in Figure 10-25, to add windows.

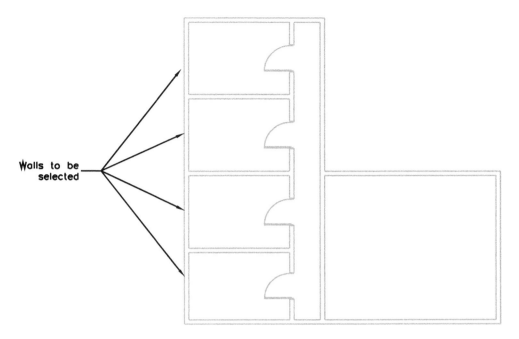

***Figure 10-25*** *The walls to be selected for adding windows*

The drawing after adding windows is displayed, as shown in Figure 10-26.

***Figure 10-26*** *The drawing after adding windows*

4.  Save the drawing and close it.

# Creating HVAC Drawings

Now, you need to create HVAC drawings for the building. For creating HVAC drawings you need to switch to **HVAC** workspace.

## Creating HVAC Drawing of the Ground Floor

1. Select the **Mechanical** category from the **Project Navigator** and choose the **Add Construct** button available at the bottom; the **Add Construct** dialog box is displayed.

2. Enter the name **Ground Floor Mechanical** in the **Name** field of the dialog box.

3. Click in the **Description** field; the **Description** dialog box is displayed.

4. Enter the description as **This drawing is meant for HVAC plan of the ground floor** and then choose the **OK** button from the dialog box.

5. Select the check box corresponding to level 0 in the **Division** column in the **Add Construct** dialog box and choose the **OK** button from the dialog box; the drawing is opened in the AutoCAD MEP. Also, the Ground Floor Mechanical drawing is displayed with a lock icon adjacent to it in the **Project Navigator**.

6. Click on the **Workspace Switching** option in the Drawing Status Bar and choose the **HVAC** option from the flyout.

## Importing the Architectural Drawing of the Ground Floor

Before adding any HVAC system in the project, you need to import the architectural plan of the ground floor.

1. Open the **Project Navigator**, and then click on the plus sign adjacent to Architectural category in the **Constructs** tab; list of architectural drawings is displayed.

2. Select Ground Floor Architectural drawing from the list and drag it to the current drawing; the architectural drawing is attached to the current drawing as an external reference.

## Adding Air Terminals in the Architectural Drawing of the Ground Floor

1. Choose the **Air Terminal** tool from the **Equipment** panel of the **Home** tab in the **Ribbon**; the **Add Multi-view Parts** dialog box is displayed, as shown in Figure 10-27.

2. Select the **600 x 600 mm Square Faced Ceiling Diffuser** part from the **Part** tab of the dialog box; the preview of the diffuser is displayed in the right of the dialog box.

3. Select the **600 x 600 mm Square Plaque Face Ceiling Diffuser -150 mm Neck** option from the **Part Size Name** drop-down list at the bottom of the dialog box.

4. Specify the value **3200** in the **Elevation** edit box.

*Figure 10-27  The **Add Multi-view Parts** dialog box*

5.  Choose the **Flow** tab; the dialog box is modified, as shown in Figure 10-28.

*Figure 10-28  The **Add Multi-view Parts** dialog box with the **Flow** tab chosen*

6.  Specify **8** in the **Flow (Each Terminal)** edit box.

7.  Place the diffusers symmetrically centered, as shown in Figure 10-29.

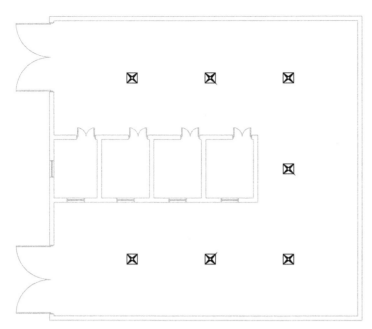

*Figure 10-29* *The drawing after adding diffusers in the outer area*

Now, you need to add the air terminals in the offices having a flow rate of 2 liter per hour.

8.  Choose the **Air Terminal** tool from the **Equipment** panel of the **Home** tab in the **Ribbon**; the **Add Multi-view Parts** dialog box is displayed.

9.  Select the **600 x 600 mm Square Faced Ceiling Diffuser** part from the **Part** tab of the dialog box; the preview of the diffuser is displayed in the right of the dialog box.

10. Choose the **600 x 600 mm Square Plaque Face Ceiling Diffuser -150 mm Neck** option from the **Part Size Name** drop-down list at the bottom of the dialog box.

11. Specify the value **3200** in the **Elevation** edit box.

12. Choose the **Flow** tab and specify the value of flow as **2** in the **Flow** edit box of the dialog box.

13. Place the diffusers symmetrically centered in the offices, refer to Figure 10-30.

## Creating the Duct Line of Ground Floor

1.  Choose **Tools** from the **Tools** drop-down in the **Build** panel of the **Home** tab in the **Ribbon**; the **Tool Palettes** is displayed.

2.  Click on the **Properties** button on the top left of the **Tool Palettes**; a flyout is displayed, as shown in Figure 10-31.

*Figure 10-30*  *The drawing after adding diffusers in the offices*

*Figure 10-31*  *The **Properties** flyout of the **Tool Palettes***

3.  Choose the **HVAC** option from the flyout; the tools related to HVAC system are displayed in the **Tool Palettes**, refer to Figure 10-32.

*Figure 10-32* The **Tool Palettes - HVAC**

4.  Choose the **1-Line** tool from the **Tool Palettes**; you are prompted to specify the start point of the duct.

5.  Click on the **Duct End Connector** of the first duct at the top in the outer area of the plant, refer to Figure 10-33.

6.  Click on the duct adjacent to the previous one to connect it to the duct line. Similarly, connect all the other ducts to the duct line, refer to Figure 10-34. To connect the ducts at the same elevation, you might need to switch to isometric views.

*Figure 10-33* The duct to be selected for adding duct line

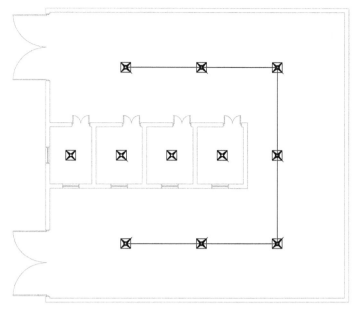

**Figure 10-34** *The duct line created between the outer ducts*

7. Connect all the ducts in the office area to the duct at the middle of the duct line using the **1-Line** tool, refer to Figure 10-35.

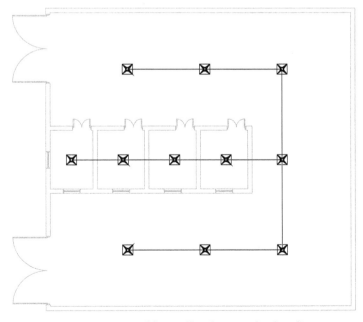

**Figure 10-35** *Adding office ducts to the duct line*

8. Similarly, connect a duct line to the main duct line for calculating the duct size and transforming the duct line into ducts of required sizes, refer to Figure 10-36.

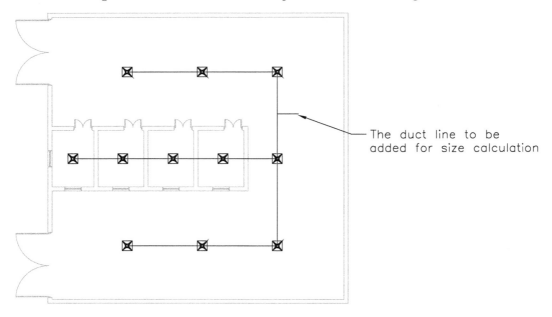

*Figure 10-36*  *Adding a duct line for load calculation*

## Calculating the Duct Sizes

1. Select the duct line created for load calculations; the **Duct** contextual tab is displayed, as shown in Figure 10-37.

*Figure 10-37*  *The **Duct** contextual tab*

2. Choose the **Calculate Duct Sizes** tool from the **Calculations** panel of the **Duct** contextual tab in the **Ribbon**; the **Duct System Size Calculator** dialog box is displayed, as shown in Figure 10-38.

3. Select the **Round** option from the first drop-down list in the **2** area of the dialog box.

4. Choose the **Start** button from the **4** area of the dialog box; the **Choose a Part** dialog box is displayed, as shown in Figure 10-39 and you are prompted to select a suitable part from the part list available at the left of the dialog box.

5.  Select the desired part from the left area of the dialog box and choose the **OK** button from the dialog box; the desired part is added and the **Choose a Part** dialog box is displayed again.

6.  Select the desired part from the dialog box and choose the **OK** button. Repeat the procedure till all the parts are added; the success rate of the duct size is displayed in the **5** area of the dialog box.

    If the success rate is not 100%, then there is an error in the fittings added in the duct line.

7.  To check the problems in the duct line, choose the **View Event Log** button from the **5** area of the dialog box; the **Event Log** dialog box is displayed, refer to Figure 10-40. In such cases, you need to rearrange the duct system. The duct line after rearranging is shown in Figure 10-41.

*Figure 10-38* The **Duct System Size Calculator** dialog box

*Figure 10-39* The **Choose a Part** dialog box

*Figure 10-40  The **Event Log** dialog box*

**Note**
*The **Choose a Part** dialog box will be displayed during the transition only when there is an elevation or size difference between the two mating ducts .*

*Figure 10-41  The Duct line after rearrangement*

8.  Again, select the duct line created for calculating duct sizes and choose the **Calculate Duct Sizes** tool from the **Duct** contextual tab; the **Duct System Size Calculator** dialog box is displayed.

9.  Choose the **Start** button from the **4** area of the dialog box and add the desired part by using the **Select a Part** dialog box; the system applies the transitions automatically and the duct system is displayed, as shown in Figure 10-42.

*Figure 10-42* *The Duct system after calculating duct sizes and applying transitions*

## Creating Return Duct for the Ground Floor

1.  Choose the **Air Terminal** tool from the **Equipment** drop-down in the **Build** panel of the **Home** tab in the **Ribbon**; the **Add Multi-view Parts** dialog box is displayed, as shown in Figure 10-43.

*Figure 10-43* *The **Add Multi-view Parts** dialog box*

2.  Click on the plus sign adjacent to **Grilles** in the **Part** tab of the dialog box; various grilles available in AutoCAD MEP are displayed.

3.  Select the **Return Air Grilles** part from the part list; the preview of the parts is displayed in the right of the dialog box.

4.  Select the **75 X 625 mm Return Air Grille** option from the **Part Size Name** drop-down list.

5.  Specify the value **3600** in the **Elevation** edit box available at the bottom.

6.  Click on the **Flow** tab and specify the value **8** in the **Flow (Each Terminal)** edit box.

7.  Add the return grilles, as shown in Figure 10-44.

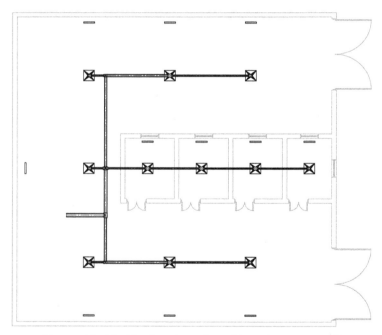

*Figure 10-44  The drawing after adding the return grilles*

## Creating Return Duct Line for the Ground Floor

1.  Choose the **1-Line** tool from the **By Shape** area of the **Tool Palettes**; you are prompted to specify the start point of the duct line.

2.  Click on the return grill and create the duct line, refer to Figure 10-45.

## Calculating the Duct Sizes for return grill system

1.  Select a duct line in the return grill system; the **Duct** contextual tab is displayed.

2. Choose the **Calculate Duct Sizes** tool from the **Calculations** panel in the contextual tab; the **Duct System Size Calculator** dialog box is displayed. Make sure that **Round** is selected in the first drop-down list of the **2** area in the dialog box.

*Figure 10-45  The drawing after connecting the return grilles with the duct line*

3. Choose the **Start** tool from the **4** area of the dialog box and choose the **OK** button from the successive dialog boxes; the duct sizes are calculated and transition is applied automatically.

4. Choose the **Close** button from the **Duct System Size Calculator** dialog box to exit.

5. Save the drawing and close it.

## Creating Mechanical Drawing of the First Floor

1. Select the **Mechanical** category from the **Project Navigator** and choose the **Add Construct** button available at the bottom; the **Add Construct** dialog box is displayed.

2. Enter the name **First Floor Mechanical** in the **Name** field of the dialog box.

3. Click in the **Description** field; the **Description** dialog box is displayed.

4. Enter the description as **This drawing is meant for mechanical plan of the first floor** and then choose the **OK** button from the dialog box.

5. Select the check box corresponding to level 1 in the **Division** column in the **Add Construct** dialog box and choose the **OK** button from the dialog box; the drawing is opened. Also,

the First Floor Mechanical drawing is displayed with a lock icon adjacent to it in the Project Navigator.

6.  Click on the **Workspace Switching** option in the **Drawing Status Bar** and choose the **HVAC** option from the flyout.

## Importing the Architectural Drawing of the Ground Floor

Before adding a mechanical system in this drawing, you need to first import the architectural plan of the first floor.

1.  Open the **Project Navigator** and then click on the plus sign adjacent to Architectural category in the **Constructs** tab; list of architectural drawings is displayed.

2.  Select First Floor Architectural drawing from the list and drag it to the current drawing; the architectural drawing is attached to the current drawing as an external reference.

## Adding Air Terminals in the Architectural Drawing of the First Floor

1.  Choose the **Air Terminal** tool from the **Equipment** panel of the **Home** tab in the **Ribbon**; the **Add Multi-view Parts** dialog box is displayed.

2.  Select the **600 x 600 mm Square Faced Ceiling Diffuser** part from the **Part** tab of the dialog box; the preview of the diffuser is displayed in the right of the dialog box.

3.  Choose the **600 x 600 mm Square Plaque Face Ceiling Diffuser -150 mm Neck** option from the **Part Size Name** drop-down list at the bottom of the dialog box.

4.  Specify the value **3200** in the **Elevation** edit box.

5.  Choose the **Flow** tab; the dialog box is modified, as shown in Figure 10-46.

*Figure 10-46  The **Add Multi-view Parts** dialog box with the **Flow** tab chosen*

6.  Specify the value as **4** in the **Flow** edit box.

7.  Place the diffusers symmetrically centered, as shown in Figure 10-47.

*Figure 10-47* *The drawing after adding diffusers*

## Adding Duct Line to the Diffusers

1.  Choose the **1-Line** tool from the **By Shape** area of the **Tool Palettes**; you are prompted to specify the start point of the duct line.

2.  Click on the Duct End Connectors and connect the ducts to the duct line, refer to Figure 10-48.

*Figure 10-48* *The drawing after adding the duct line*

## Adding Air Handling Unit

1. Choose the **Air Handler** tool from the **Equipment** drop-down in the **Build** panel of the **Home** tab in the **Ribbon**; the **Add Multi-view Parts** dialog box is displayed, as shown in Figure 10-49. Also, you are prompted to specify the insertion point for the air handler unit.

*Figure 10-49*  The **Add Multi-view Parts** dialog box

2. Click on the plus sign adjacent to **Packaged Air Handling Unit** in the part list; the list of air handling units is displayed.

3. Select **Air Handling Unit - Floor Mounted Front Discharge** from the list; preview of the air handling unit is displayed on the right in the dialog box.

4. Set the value of elevation as **0** in the **Elevation** edit box available at the bottom of the dialog box and click in the drawing area to place the air handling unit, refer to Figure 10-50.

*Figure 10-50*  The drawing after placing the air handling unit

5.   Connect the diffusers with the air handling unit, refer to Figure 10-51.

*Figure 10-51  The drawing after connecting the diffusers to the air handling unit*

## Adding Return Grilles in the Architectural Drawing of First Floor

1.   Choose the **Air Terminal** tool from the **Equipment** panel of the **Home** tab in the **Ribbon**; the **Add Multi-view Parts** dialog box is displayed. Also, you are prompted to specify the insertion point of the air terminal.

2.   Click on the plus sign adjacent to **Grilles** in the part list and then choose the **Return Air Grilles** part from the part list; the preview of the air grilles is displayed in the right of the dialog box, refer to Figure 10-52.

*Figure 10-52  The **Add Multi-view Parts** dialog box*

3.  Select the **75 x 625 mm Return Air Grille** option from the **Part Size Name** drop-down list and specify the value of elevation **3600** in the **Elevation** edit box.

4.  Click on the **Flow** tab and specify the value for flow as **4** in the **Flow (Each Terminal)** edit box.

5.  Add the return grilles, as shown in Figure 10-53.

*Figure 10-53 The drawing after adding return grilles*

## Creating Return Duct Line for the First Floor

1.  Choose the **1-Line** tool from the **By Shape** area of the **Tool Palettes**; you are prompted to specify the start point of the duct line.

2.  Click on the return grill and create the duct line, refer to Figure 10-54.

## Calculating the Duct Sizes

1.  Select a duct line in the diffuser system; the **Duct** contextual tab is displayed.

2.  Choose the **Calculate Duct Sizes** tool from the **Calculations** panel in the tab; the **Duct System Size Calculator** dialog box is displayed.

3.  Choose the **Start** button from the **4** area of the dialog box; the **Choose a Part** dialog box is displayed.

4.  Choose the **OK** button from the successive **Choose a Part** dialog boxes and then choose the **Close** button from the **Duct System Size Calculator** dialog box; the duct line after applying transitions is displayed, refer to Figure 10-55.

*Figure 10-54* *The drawing after adding duct line to the return grilles*

*Figure 10-55* *The drawing after applying transitions to the diffuser duct line*

5.  Similarly, calculate the duct sizes of the duct line created for return duct. On doing so, the drawing is displayed, as shown in Figure 10-56.

*Figure 10-56  The drawing after applying transitions to the return grille duct line*

6.  Save the drawing and close it.

## Creating Piping Drawings

Now, you need to create piping drawings in the building. To create piping drawings you need to switch to the **Piping** Workspace.

### Creating Piping Drawing of the Ground Floor

1.  Select the **Piping** category from the **Project Navigator** and choose the **Add Construct** button available at the bottom; the **Add Construct** dialog box is displayed.

2.  Enter the name **Ground Floor Piping** in the **Name** field of the dialog box.

3.  Click in the **Description** field; the **Description** dialog box is displayed.

4.  Enter the description as **This drawing is meant for piping plan of the ground floor** and then choose the **OK** button from the dialog box.

5.  Select the check box corresponding to level 0 in the **Division** column in the **Add Construct** dialog box and choose the **OK** button from the dialog box; the drawing is opened. Also, the Ground Floor Piping drawing is displayed with a lock icon adjacent to it in the Project Navigator.

6. Click on the **Workspace Switching** option in the Drawing Status Bar and choose the **Piping** option from the flyout.

## Importing the Architectural Drawing of the Ground Floor

Before adding any mechanical system in the project, you need to first import the architectural plan of the ground floor.

1. Open the **Project Navigator** and then click on the plus sign adjacent to Architectural category in the **Constructs** tab; a list of architectural drawings is displayed.

2. Select Ground Floor Architectural drawing from the list and drag it into the current drawing; the architectural drawing is attached to the current drawing as an external reference.

## Adding Ball Valves in the Architectural Drawing of the Ground Floor

1. Choose the **Valve** tool from the **Equipment** panel of the **Home** tab in the **Ribbon**; the **Add Multi-view Parts** dialog box is displayed, as shown in Figure 10-57.

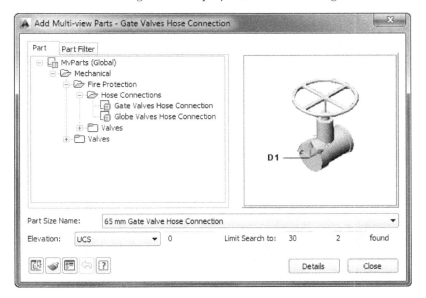

*Figure 10-57 The **Add Multi-view Parts** dialog box*

2. Click on the plus sign adjacent to **Valves** in the part list of the dialog box; various types of valves are displayed in the list.

3. Select the **Ball Valve - Threaded - Nickel plated** part from the list; the preview of the part is displayed in the right of the dialog box, refer to Figure 10-58.

*Figure 10-58* *The **Add Multi-view Parts** dialog box with the ball valve selected*

4.  Select the **40 mm Ball valve - Threaded - Nickel plated ('A' = 108 'B' = 161 'C' = 80)** option from the **Part Size Name** drop-down list.

5.  Specify the elevation value as **1000** in the **Elevation** edit box.

6.  Place the ball valves symmetrically, as shown in Figure 10-59.

*Figure 10-59* *The drawing after adding ball valves*

## Adding Pipe Line in the Architectural Drawing of Ground Floor

1. Choose the **Pipe** tool from the **Pipe** drop-down in the **Build** panel of the **Home** tab in the **Ribbon**; the **Properties** palette is displayed, as shown in Figure 10-60. Also, you are prompted to specify the start point of the pipe.

2. Click in the **Nominal Size** drop-down list and select the **40** option from the drop-down list.

3. Click in the **Specify cut length** drop-down list and select the **Yes** option from the drop-down list; the **Cut length** edit box is displayed below the drop-down list.

4. Specify **6000** in the **Cut length** edit box.

5. Click in the drawing area and draw the pipe line, as shown in Figure 10-61.

## Connecting Ball Valves with the Pipe Line

1. Click on the ball valve at the left end of the bottom row; the valve is displayed with plus signs, refer to Figure 10-62.

*Figure 10-60  The **Properties** palette on choosing the **Pipe** tool*

*Figure 10-61  The drawing after drawing the pipe line*

***Figure 10-62***  *The ball valve with plus signs*

2.  Click on the plus sign, refer to Figure 10-62 and connect the ball valve with the pipe line, refer to Figure 10-63.

3.  Similarly, connect all the ball valves to the pipe line, refer to Figure 10-64.

***Figure 10-63***  *The ball valve after connecting to the pipe line*

*Figure 10-64* *Piping system after connecting all the ball valves to the pipe line*

4.  Select the elbow at the bottom left corner of the drawing and then click on the plus sign displayed on the elbow, as shown in Figure 10-65; the elbow transforms into a tee.

*Figure 10-65* *The plus sign to be selected*

5.  Move the cursor horizontally towards the left and specify **1000** at the command prompt. Press ENTER again to exit the tool.

6.  Save the drawing and close it.

## Creating Piping Drawing of the First Floor

1.  Select the **Piping** category from the **Project Navigator** and choose the **Add Construct** button available at the bottom; the **Add Construct** dialog box is displayed.

2.  Enter **First Floor Piping** in the **Name** field of the dialog box.

3.  Click in the **Description** field; the **Description** dialog box is displayed.

4.  Enter the description **This drawing is meant for piping plan of the first floor** in the **Edit the Description** edit box and then choose the **OK** button from the dialog box.

5.  Select the check box corresponding to level 1 in the **Division** column of the **Add Construct** dialog box and then choose the **OK** button from the dialog box; the drawing is opened in AutoCAD MEP. Also, the First Floor Piping drawing is displayed with a lock icon adjacent to it in the Project Navigator.

6.  Click on the **Workspace Switching** option in the Drawing Status Bar and then choose the **Piping** option from the flyout; the **Piping** workspace gets activated.

## Importing the Architectural Drawing of the First Floor

Before adding any piping system to the project, first you need to import the architectural plan of the first floor.

1.  Open the **Project Navigator** and then click on the plus sign adjacent to Architectural category in the **Constructs** tab; a list of architectural drawings is displayed.

2.  Select the First Floor Architectural drawing from the list and drag it to the current drawing; the architectural drawing is attached to the current drawing as an external reference.

## Adding Tank in the First Floor

1.  Choose the **Tank** tool from the **Equipment** panel of the **Home** tab in the **Ribbon**; the **Add Multi-view Parts** dialog box is displayed, as shown in Figure 10-66.

2.  Select the **Horizontal Storage Tanks** part from the part list; the preview of tank is displayed on the right in the dialog box.

3.  Select the **1300 Liter Horizontal Storage Tank** option from the **Part Size Name** drop-down list and specify the value of elevation as **0** in the **Elevation** edit box.

4.  Click in the drawing area to specify the insertion point of the tank, refer to Figure 10-67.

*Figure 10-66  The **Add Multi-view Parts** dialog box*

*Figure 10-67* *The drawing after adding the tank*

## Adding Pumps in the First Floor

1. Choose the **Pumps** tool from the **Equipment** panel of the **Home** tab in the **Ribbon**; the **Add Multi-view Parts** dialog box is displayed.

2. Select the **Horizontal Split Case Pumps** part from the part list available at the left of the dialog box; the preview of the pump is displayed in the right of the dialog box.

3. Select the **1200 x 600 mm Horizontal Split Case** pump option from the **Part Size Name** drop-down list and specify the value of elevation as **0** in the **Elevation** edit box available at the bottom of the dialog box.

4. Click in the drawing area to specify the insertion point for the pump, refer to Figure 10-68.

*Figure 10-68* *The drawing after adding pump*

## Adding Pipe Line in the First Floor

1.  Select the tank in the drawing area; plus signs are displayed on the tank.

2.  Click on the plus sign at the right of the tank; a rubber band pipe gets attached to the cursor, refer to Figure 10-69.

3.  Click on the adjacent Pipe End Connector of the pump; the **Choose a Part** dialog box is displayed, as shown in Figure 10-70.

*Figure 10-69  The rubber band pipe attached to the cursor*

*Figure 10-70  The **Choose a Part** dialog box*

4. Choose the **OK** button from the dialog boxes displayed successively and then press ENTER; a pipe line is created between the tank and the pump.

5. Similarly, create the pipe line, as shown in Figure 10-71.

6. Save the drawing and close it.

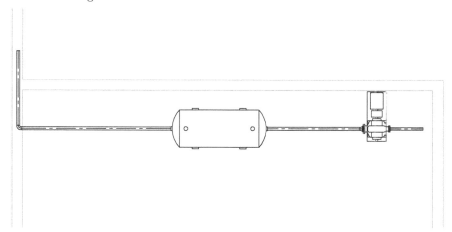

*Figure 10-71  The drawing after adding pipe lines*

## Creating Electrical Drawings

Now, you need to create electrical drawings of the building. To do so, you need to switch to the **Electrical** workspace.

### Creating Electrical Drawing of the Ground Floor

1. Select the **Electrical** category from the **Project Navigator** and choose the **Add Construct** button available at the bottom; the **Add Construct** dialog box is displayed.

2. Enter **Ground Floor Electrical** in the **Name** field of the dialog box.

3. Click in the **Description** field; the **Description** dialog box is displayed.

4. Enter the description **This drawing is meant for electrical plan of the ground floor** in the **Edit the Description** edit box and then choose the **OK** button from the dialog box.

5. Select the check box corresponding to level 0 in the **Division** column in the **Add Construct** dialog box and then choose the **OK** button from the dialog box; the drawing is opened in AutoCAD MEP. Also, the Ground Floor Electrical drawing is displayed with a lock icon adjacent to it in the **Project Navigator**.

6. Click on the **Workspace Switching** option in the Drawing Status Bar and select the **Electrical** option from the flyout.

## Importing the Architectural Drawing of the Ground Floor

Before adding any electrical system in the project, you need to import the architectural plan of the ground floor.

1.  Open the **Project Navigator** and then click on the plus sign adjacent to Architectural category in the **Constructs** tab; a list of architectural drawings is displayed.

2.  Select the Ground Floor Architectural drawing from the list and drag it to the current drawing; the architectural drawing is attached to the current drawing as an external reference.

## Adding 300X1200 Surface Type A Lights in the Drawing

Now, you need to add lights to the drawing.

1.  Choose the **Device** tool from the **Build** panel of the **Home** tab in the **Ribbon**; the **Properties** palette is modified, refer to Figure 10-72.

*Figure 10-72  The **Properties** palette displayed on choosing the **Device** tool*

2.  Click in the **Style** field of the **Properties** palette; the **Select a Style** dialog box is displayed.

3.  Select the **Lighting Fluorescent (Global)** option from the **Drawing file** drop-down list; the light styles available in AutoCAD MEP are displayed.

4.  Select the **300X1200 Surface Type A Light** from the light styles and then choose the **OK** button from the dialog box; the light gets attached to the cursor.

5.  Place the lights in the drawing, as shown in Figure 10-73.

*Figure 10-73* *The drawing after adding lights*

## Adding Sockets in the Drawing
Now, you need to add sockets to the drawing.

1.  Choose the **Device** tool from the **Build** panel; the **Properties** palette is displayed.

2.  Click in the **Style** field of the **Properties** palette; the **Select a Style** dialog box is displayed.

3.  Select the **Sockets (Global)** option from the **Drawing file** drop-down list; various styles of sockets available in AutoCAD MEP are displayed in the dialog box.

4.  Select the **Single Pole Switched Outlet** socket style from the dialog box and then choose the **OK** button; the socket gets attached to the cursor.

5.  Place the sockets along the wall, as shown in Figure 10-74.

## Adding 300x600 Recessed Lights in the Drawing
Now, you need to add lights in the drawing.

1.  Choose the **Device** tool from the **Build** panel; the **Properties** palette is displayed.

*Figure 10-74* *The drawing after adding sockets*

2.  Click in the **Style** field of the **Properties** palette; the **Select a Style** dialog box is displayed.

3.  Select the **Lighting Fluorescent (Global)** option from the **Drawing file** drop-down list; various styles of lights available in AutoCAD MEP are displayed in the dialog box.

4.  Select the **300x600 Recessed Lights** style from the dialog box and then choose the **OK** button; the light gets attached to the cursor.

5.  Place the lights in the center of inner area of the ground floor, refer to Figure 10-75.

## Configuring Devices

1.  Select a socket from the drawing area and then select the **Select Similar** option from the **Select Similar** drop-down list in the **Device** contextual tab displayed; all the sockets available in the drawing area are selected.

2.  Click on the **Electrical properties** field in the **Advanced** rollout of the **Properties** palette; the **Electrical Properties** dialog box is displayed, as shown in Figure 10-76.

*Figure 10-75* *The drawing after adding lights in the center of the inner area*

*Figure 10-76* *The **Electrical Properties** dialog box*

Now, you need to specify the parameters that are given in Table 10-1.

3.  Select the **230V Power** option from the **System** drop-down list.

4.  Specify the value **300** in the **Load Phase 1** edit box, refer to Table 10-1.

5.  Select the **230** option from the **Voltage** drop-down list.

6.  Similarly, select the **1** option from the **Number of Poles** drop-down list in the dialog box.

7.  Enter the value of maximum overcurrent rating as **10** in the **Maximum Overcurrent Rating (amps)** edit box in the dialog box and.

8.  Specify the value as **0.8** in the **Power Factor** edit box. Choose the **OK** button to exit the dialog box.

9.  Similarly, specify the parameters for other devices in the **Electrical Properties** dialog box by using the steps discussed above. For parameters, refer to Table 10-1.

## Creating Panel

All the devices are connected to circuits. These circuits are joined to a panel for electricity supply. Therefore, you need to create a panel with circuits in this section.

1.  Choose the **Panel** tool from the **Build** panel in the **Home** tab of the **Ribbon**; you are prompted to specify an insertion point for the panel and the **Properties** palette is displayed, as shown in Figure 10-77.

2.  Click in the **Name** edit box in the **ADVANCED > Design Data** rollout of the **Properties** palette and specify the name as **Main Panel**.

3.  Specify the value **800** in the **Rating** edit box.

4.  Select the **230** option from the **Voltage phase-to-neutral** drop-down list.

5.  Select the **240** option from the **Voltage phase-to-phase** drop-down list.

6.  Specify **Main type** as **Main circuit breaker**, **Main size (amps)** as **15**, **Design capacity (amps)** as **20**, and **AIC rating** as **800** in the **Properties** palette.

*Figure 10-77  The **Properties** palette displayed on choosing the **Panel** tool*

Now, you need to create circuits for the panel.

7. Click in the **Circuit Settings** field of the **Properties** palette; the **Circuit Settings** dialog box is displayed, as shown in Figure 10-78.

*Figure 10-78 The **Circuit Settings** dialog box*

8. Set the value as **2** in the **Total number of slots** and **Number of 1-pole circuits** spinners and select the **230** option from the **Voltage** drop-down list adjacent to the spinner. Make sure that **Power and Lighting** is selected in the **System Type** drop-down list and **230V Power** is selected in the **System** drop-down list.

9. Choose the **OK** button from the dialog box to exit.

10. Select the **Yes** option from the **Create Circuit** drop-down list in the **Advanced** rollout. Click in the drawing area to place the panel, refer to Figure 10-79. Also, the **AutoCAD MEP - Electrical Project Database** dialog box is displayed, refer to Figure 10-80.

11. Choose the **Create a new EPD file** option from the dialog box and save the file at the desired location with the name **light_panel**; the panel is created at the specified location.

*Figure 10-79  The drawing with the panel placed*

*Figure 10-80  The **AutoCAD MEP - Electrical Project Database** dialog box*

12. Similarly, place the other panel for 230V Power (230V POWER) system, refer to Figure 10-81.

*Figure 10-81* *The drawing with the other panel added*

## Configuring Circuits

1. Select the panel that was created first and then choose the **Circuit Manager** tool from the **Circuits** panel in the **Panel** contextual tab of the **Ribbon**; the **Circuit Manager** is displayed, refer to Figure 10-82.

*Figure 10-82* *The Circuit Manager*

2. Specify the name **Outer Lights** in the field **1** of the **Name** column in the **Circuit Manager**.

3. Specify the name **Inner Lights** in the field **2** of the **Name** column in the **Circuit Manager**.

4. Click on the plus sign adjacent to **Sockets panel** on the left in the **Circuit Manager**; the list of circuits available in the panel is displayed below it.

5. Select **Power and Lighting** from the left of the **Circuit Manager**; the list of circuits is displayed on the right.

6. Double-click on **1** under the **Name** panel and specify the name as **Sockets**.

7. Double-click in the **Voltage** field for **Sockets** and select the **230** option from the drop-down list.

8. Select the **230V Power** option from the **System** drop-down list.

## Adding Wires

1. Choose the **Wire** tool from the **Build** panel of the **Home** tab in the **Ribbon**; you are prompted to specify the start point of the wire for an electrical device and the **Properties** palette is displayed, as shown in Figure 10-83.

*Figure 10-83   The **Properties** palette displayed on choosing the **Wire** tool*

2. Select the **230V Power(230V POWER)** option from the **System** drop-down list.

3. Select the **PVC Single** option from the **Style** drop-down list.

4. Connect all the sockets using wires, refer to Figure 10-84. Make sure that the **Line** option is selected in the **Segment** drop-down list.

***Figure 10-84*** *The drawing after connecting sockets using wires*

5. Similarly, connect all the lights to the other panel using the 230V Lighting system, refer to Figure 10-85. Make sure that the circuit selected for inner lights is **Inner Lights** and for outer lights is **Outer Lights**.

***Figure 10-85*** *The drawing after adding wires to lights*

## Calculating Load and Wire Sizes

Now, you need to calculate the total load of the electrical system and the wire size.

1.  Choose the **Power Totals** tool from the **Electrical** panel in the **Analyze** tab of the **Ribbon**; you are prompted to select the devices.

2.  Select all the devices available in the drawing area and press ENTER; the **Power Totals** dialog box is displayed, refer to Figure 10-86.

*Figure 10-86  The Power Totals dialog box*

The total load is displayed in the **Total Load** field of the dialog box. Now, you need to calculate the wire size for the circuit.

3.  To calculate wire size, select all the wires in the drawing area; the **Properties** palette is displayed.

4.  Click on the **Calculate sizes for the wire** button available on the right of the **ADVANCED > Dimensions** rollout of the **Properties** palette; the wire sizes are displayed in the **Hot size**, **Neutral size**, and **Ground size** edit boxes in the **Dimensions** rollout.

# Adding Transformer and Emergency Generator to the Drawing Based on Total Load

After making calculations the total load of the ground floor is approximately 48kVa. Assuming the same capacity for first floor and considering the peak load condition, you need to place a transformer and a generator of capacity 100kVa.

1.  Choose the **Equipment** tool from the **Equipment** drop-down in the **Build** panel of the **Home** tab in the **Ribbon**; the **Add Multi-view Parts** dialog box is displayed, refer to Figure 10-87.

2.  Click on the plus sign adjacent to **Electrical** in the part list available at the left of the dialog box; the electrical equipment are displayed below it.

3.  Click on the plus sign adjacent to **Power Transformers** in the part list and select the **Dry Type Transformer - 3-150 kVa** part; the preview of the transformer is displayed, refer to Figure 10-87.

4.  Select the **112 kVa Dry Type Transformer** option from the **Part Size Name** drop-down list available at the bottom of the dialog box and place the transformer at an appropriate distance from the building in the outer area, refer to Figure 10-88.

*Figure 10-87* The **Add Multi-view Parts** *dialog box*

*Figure 10-88* The drawing after placing a transformer

5.  Similarly, place the emergency generator having the corresponding capacity in the drawing adjacent to the power transformer.

6.  Connect the transformer and emergency generator with the panels, refer to Figure 10-89.

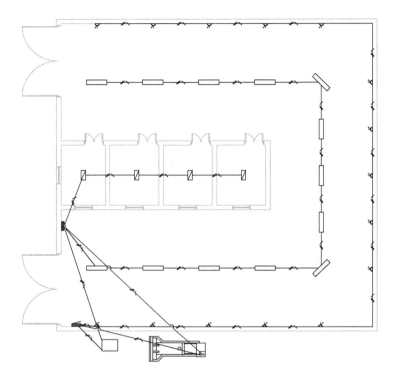

*Figure 10-89* *The drawing after connecting transformer and generator to the circuit*

## Creating Electrical Drawing of the First Floor

1.  Select the **Electrical** category from the **Project Navigator** and choose the **Add Construct** button available at the bottom; the **Add Construct** dialog box is displayed.

2.  Click in the **Name** field of the dialog box and enter the name as **First Floor Electrical**.

3.  Click in the **Description** field; the **Description** dialog box is displayed.

4.  Enter the description **This drawing is meant for electrical plan of the first floor** and then choose the **OK** button from the dialog box.

5.  Select the check box corresponding to level 1 in the **Division** column of the **Add Construct** dialog box and choose the **OK** button from the dialog box; the drawing is opened in AutoCAD MEP. Also, the First Floor Electrical drawing is displayed with a lock icon adjacent to it in the Project Navigator.

6.  Click on the **Workspace Switching** option of the Drawing Status Bar and choose the **Electrical** option from the flyout.

## Importing the Architectural Drawing of the First Floor

Before adding any mechanical system to the project, you need to first import the architectural plan of the first floor.

1. Open the **Project Navigator** and then click on the plus sign adjacent to Architectural category in the **Constructs** tab; the list of architectural drawings is displayed.

2. Select First Floor Architectural drawing from the list and drag it to the current drawing; the architectural drawing is attached to the current drawing as an external reference.

## Adding Lights in the Drawing

Now, you need to add lights to the drawing.

1. Choose the **Device** tool from the **Build** panel of the **Home** tab in the **Ribbon**; the **Properties** palette is displayed, refer to Figure 10-90.

2. Click in the **Style** field of the **Properties** palette; the **Select a Style** dialog box is displayed.

3. Select the **Lighting Fluorescent (Global)** option from the **Drawing file** drop-down list; the light styles available in AutoCAD MEP are displayed.

*Figure 10-90  The **Properties** palette displayed on choosing the **Device** tool*

4. Select the **300X600 Surface Type A Light** from the light styles and then choose the **OK** button from the dialog box; the light symbol gets attached to the cursor.

5. Place three equidistant lights in each room in the drawing, refer to Figure 10-91.

## Adding Sockets in the Drawing

Now, you need to add sockets to the drawing.

1. Choose the **Device** tool from the **Build** panel; the **Properties** palette is displayed.

2. Click in the **Style** field of the **Properties** palette; the **Select a Style** dialog box is displayed.

3. Select the **Sockets (Global)** option from the **Drawing file** drop-down list; various styles of sockets available in AutoCAD MEP are displayed in the dialog box.

4. Select the **Single Pole Switched Outlet** socket style from the dialog box and then choose the **OK** button; the socket symbol gets attached to the cursor.

5. Place the sockets along the wall, refer to Figure 10-91.

*Figure 10-91* The drawing after adding lights and sockets

## Configuring Devices

1. Select a socket from the drawing area and then choose the **Select Similar** option from the **Select Similar** drop-down list in the **Device** contextual tab displayed; all the sockets available in the drawing area are selected.

2. Click on the **Electrical properties** field in the **Circuit** rollout of the **Properties** palette; the **Electrical Properties** dialog box is displayed, as shown in Figure 10-92.

   Now, you need to specify the parameters that are given in Table 10-1.

3. Choose the **230V Power** option from the **System** drop-down list.

4. Specify the value **300** in the **Load Phase 1** edit box, refer to Table 10-1.

5. Select the **230** option from the **Voltage** drop-down list.

6. Similarly, select the **1** option from the **Number of Poles** drop-down list in the dialog box.

7. Specify the value of maximum overcurrent rating **10** in the **Maximum Overcurrent Rating (amps)** edit box in the dialog box.

8. Specify the value as **0.8** in the **Power Factor** edit box and then choose the **OK** button to exit the dialog box.

9. Similarly, specify the parameters for other devices in the **Electrical Properties** dialog box by using the steps discussed above. For parameters, refer to Table 9-1.

*Figure 10-92  The **Electrical Properties** dialog box*

## Creating Panel

As discussed earlier, all the devices are connected to circuits. These circuits are joined to a panel for electricity supply. So, you need to create a panel with circuits in this section.

1. Choose the **Panel** tool from the **Build** panel in the **Home** tab of the **Ribbon**; you are prompted to specify an insertion point for the panel and the **Properties** palette is displayed, as shown in Figure 10-93. Select the **Surface Door 3** style from the **Style** field in the **Properties** palette.

2. Click in the **Name** edit box of the **ADVANCED > Design Data** rollout of the **Properties** palette and specify the name as **Panel1**.

3. Specify the value **800** in the **Rating** edit box.

4. Select the **230** option from the **Voltage phase-to-neutral** drop-down list.

5.  Select the **240** option from the **Voltage phase-to-phase** drop-down list.

6.  Specify **Main type** as **Main circuit breaker**, **Main size (amps)** as **15**, **Design capacity (amps)** as **20**, and **AIC rating** as **800** in the **Properties** palette.

    Now, you need to create circuits for the panel.

7.  Click in the **Circuit Settings** field in the **Properties** palette; the **Circuit Settings** dialog box is displayed, as shown in Figure 10-94.

8.  Set the value as **2** in the **Total number of slots** and **Number of 1-pole circuits** spinners and select the **230** option from the **Voltage** drop-down list adjacent to the spinner. Make sure that **Power and Lighting** is selected in the **System Type** drop-down list and **230V Power** is selected in the **System** drop-down list.

*Figure 10-93* The **Properties** palette

9.  Choose the **OK** button from the dialog box to exit. Select the **Yes** option from the **Create circuits** drop-down list in the **Circuits** rollout.

*Figure 10-94* The **Circuit Settings** dialog box

10. Click in the drawing area to place the panel, refer to Figure 10-95. Also, the **AutoCAD MEP - Electrical Project Database** dialog box is displayed, refer to Figure 10-96.

*Figure 10-95  The drawing with the panel to be added*

*Figure 10-96  The AutoCAD MEP - Electrical Project Database dialog box*

11. Choose the **Create a new EPD file** option from the dialog box and save the file at the desired location with the name **top_panel**; the panel is created at the specified location.

## Configuring Circuits

1. Select the panel created and then choose the **Circuit Manager** tool from the **Circuits** panel in the **Panel** tab of the **Ribbon**; the **Circuit Manager** is displayed.

2. Click in the field **1** in the **Name** column of the **Circuit Manager** and specify the name as **Sockets**.

3. Click in the field **2** in the **Name** column of the **Circuit Manager** and specify the name as **Lights**.

4. Double click in the **Voltage** field for **Sockets** and select the **230** option from the drop-down list displayed.

5. Double click in the **System** field for the circuit and select the **230V Power** option from the drop-down list displayed.

6. Double click in the **System** field for **Lights** and select the **230V Lighting** option from the drop-down list displayed.

7. Double click in the **Voltage** field for **Lights** and select the **230** option from the drop-down list displayed.

   The **Circuit Manager**, after applying the above configuration is displayed, as shown in Figure 10-97.

*Figure 10-97* *The **Circuit Manager** after applying configuration*

## Adding Wires

1. Choose the **Wire** tool from the **Build** panel of the **Home** tab in the **Ribbon**; you are

prompted to specify the start point of the wire on an electrical device and the **Properties** palette is displayed, as shown in Figure 10-98.

2.   Select the **230V Power(230V POWER)** option from the **System** drop-down list.

3.   Select the **PVC Single** option from the **Style** drop-down list.

4.   Connect all the sockets to the panel using wire, refer to Figure 10-99. Make sure that the **Line** option is selected in the **Segment** drop-down list.

**Figure 10-98**  *The* **Properties** *palette displayed on choosing the* **Wire** *tool*

*Figure 10-99* *The drawing after adding all the sockets with the panel*

5. Similarly, connect all the lights to the panel using the 230V Lighting system, refer to Figure 10-100. Make sure that the circuit selected for lights is **Lights**.

## Calculating Loads and Wire Sizes

1. Choose the **Power Totals** tool from the **Electrical** panel in the **Analyze** tab of the **Ribbon**; you are prompted to select the devices.

2. Select all the devices available in the drawing area and press ENTER; the **Power Totals** dialog box is displayed, refer to Figure 10-101.

   The total load is displayed in the **Total Load** field of the dialog box. Now, you need to calculate the wire size for the circuit.

3. To calculate wire size, select all the wires in the drawing area; the **Properties** palette is displayed.

4.  Click on the **Calculate sizes for the wire** button available at the right of the **ADVANCED > Dimensions** rollout of the **Properties** palette; the wire sizes are displayed in the **Hot size**, **Neutral size**, and **Ground size** edit boxes in the **Dimensions** rollout.

5.  Save and close the drawing.

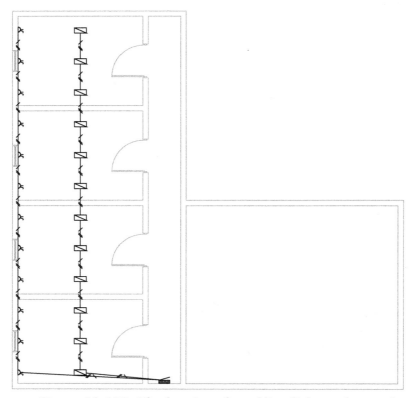

*Figure 10-100  The drawing after adding lights to the panel*

*Figure 10-101  The Power Totals dialog box*

# Index

# Other Publications by CADCIM Technologies

The following is the list of some of the publications by CADCIM Technologies. Please visit www.cadcim.com for the complete listing.

## Autodesk Inventor Textbooks
- Autodesk Inventor 2013 for Designers
- Autodesk Inventor 2012 for Designers
- Autodesk Inventor 2011 for Designers
- Autodesk Inventor 11 for Designers

## Solid Edge Textbooks
- Solid Edge ST5 for Designers
- Solid Edge ST4 for Designers
- Solid Edge ST3 for Designers
- Solid Edge ST2 for Designers
- Solid Edge ST for Designers
- Solid Edge V20 for Designers
- Solid Edge V19 for Designers
- Solid Edge V18 for Designers
- Solid Edge for Designers, Version 16

## NX Textbooks
- NX 8.5 for Designers
- NX 8 for Designers
- NX 7 for Designers
- NX 6 for Designers
- NX 5 for Designers
- NX 4 for Designers

## SolidWorks Textbooks
- SolidWorks 2013 for Designers
- SolidWorks 2012: A Tutorial Approach
- Learning SolidWorks 2011: A Project based Approach
- SolidWorks 2012 for Designers
- SolidWorks 2011 for Designers
- SolidWorks 2010 for Designers
- SolidWorks 2009 for Designers

## EdgeCAM Textbooks
• EdgeCAM 11.0 for Manufacturers
• EdgeCAM 10.0 for Manufacturers

## CATIA Textbooks
• CATIA V5-6R2012 for Designers
• CATIA V5R21 for Designers
• CATIA V5R20 for Designers
• CATIA V5R19 for Designers
• CATIA V5R18 for Designers
• CATIA V5R17 for Designers
• CATIA V5R16 for Designers
• CATIA V5R15 for Designers

## Creo Parametric and Pro/ENGINEER Textbooks
• Creo Parametric 2.0 for Designers
• Creo Parametric 1.0 for Designers
• Pro/ENGINEER Wildfire 5.0 for Designers
• Pro/ENGINEER Wildfire 4.0 for Designers
• Pro/ENGINEER Wildfire 3.0 for Designers
• Pro/ENGINEER Wildfire for Designers Release 2.0
• Pro/ENGINEER Wildfire for Designers

## Autodesk Alias Textbooks
• Learning Autodesk Alias Design 2012
• Learning Autodesk Alias Design 2010
• AliasStudio 2009 for Designers

## ANSYS Textbooks
• ANSYS Workbench 14.0: A Tutorial Approach
• ANSYS 11.0 for Designers

## Customizing AutoCAD Textbooks
• Customizing AutoCAD 2013

## AutoCAD LT Textbooks
• AutoCAD LT 2013 for Designers
• AutoCAD LT 2012 for Designers
• AutoCAD LT 2011 for Designers
• AutoCAD LT 2010 for Designers

- AutoCAD LT 2009 for Designers
- AutoCAD LT 2008 for Designers
- AutoCAD LT 2007 for Designers

## AutoCAD Electrical Textbooks
- AutoCAD Electrical 2013 for Electrical Control Designers
- AutoCAD Electrical 2012 for Electrical Control Designers
- AutoCAD Electrical 2011 for Electrical Control Designers
- AutoCAD Electrical 2010 for Electrical Control Designers

## Autodesk Revit Architecture Textbooks
- Autodesk Revit Architecture 2013 for Architects and Designers
- Autodesk Revit Architecture 2012 for Architects and Designers
- Autodesk Revit Architecture 2011 for Architects and Designers
- Autodesk Revit Architecture 2010 for Architects & Designers
- Autodesk Revit Architecture 2009 for Architects & Designers

## Autodesk Revit Structure Textbooks
- Exploring Autodesk Revit Structure 2014
- Exploring Autodesk Revit Structure 2013

## AutoCAD Civil 3D Textbooks
- Exploring AutoCAD Civil 3D 2013
- Exploring AutoCAD Civil 3D 2012
- AutoCAD Civil 3D 2009 for Engineers

## AutoCAD Map 3D Textbooks
- Exploring AutoCAD Map 3D 2013
- Exploring AutoCAD Map 3D 2012
- Exploring AutoCAD Map 3D 2011

## 3ds Max Design Textbooks
- Autodesk 3ds Max Design 2013: A Tutorial Approach
- Autodesk 3ds Max Design 2012: A Tutorial Approach
- Autodesk 3ds Max Design 2011: A Tutorial Approach
- Autodesk 3ds Max Design 2010: A Tutorial Approach
- Max Design 2009: A Tutorial Approach

## 3ds Max Textbooks
- Autodesk 3ds Max 2013: A Comprehensive Guide

- Autodesk 3ds Max 2012: A Comprehensive Guide
- Autodesk 3ds Max 2011: A Comprehensive Guide
- Autodesk 3ds Max 2010: A Comprehensive Guide
- 3ds Max 2008: A Comprehensive Guide

## Maya Textbooks
- Autodesk Maya 2013: A Comprehensive Guide
- Autodesk Maya 2012: A Comprehensive Guide
- Autodesk Maya 2011: A Comprehensive Guide
- Autodesk Maya 2010: A Comprehensive Guide
- Autodesk Maya 2009: A Comprehensive Guide
- Character Animation: A Tutorial Approach

## Fusion Textbook
- The eyeon Fusion 6.3: A Tutorial Approach

## Flash Textbook
- Adobe Flash Professional CS6: A Tutorial Approach

## Premiere Textbooks
- Adobe Premiere Pro CS6: A Tutorial Approach
- Adobe Premiere Pro CS5.5: A Tutorial Approach

## Computer Programming Textbooks
- Learning Oracle 11g: A PL/SQL Approach
- Learning ASP.NET AJAX
- Learning Java Programming
- Learning Visual Basic.NET 2008
- Learning C++ Programming Concepts
- Learning VB.NET Programming Concepts

## Paper Craft Book
- Constructing 3-Dimensional Models: A Paper-Craft Workbook

## AutoCAD Textbooks Authored by Prof. Sham Tickoo and Published by Autodesk Press
- AutoCAD: A Problem-Solving Approach: 2013 and Beyond
- AutoCAD 2012: A Problem-Solving Approach
- AutoCAD 2011: A Problem-Solving Approach

- AutoCAD 2010: A Problem-Solving Approach
- Customizing AutoCAD 2010
- AutoCAD 2009: A Problem-Solving Approach
- Customizing AutoCAD 2009
- AutoCAD 2008: A Problem-Solving Approach
- Customizing AutoCAD 2008
- AutoCAD 2007: A Problem-Solving Approach
- Customizing AutoCAD 2007

## Textbooks Authored by CADCIM Technologies and Published by Other Publishers

### 3D Studio MAX and VIZ Textbooks
- Learning 3ds max5: A Tutorial Approach
- Learning 3dS Max: A Tutorial Approach, Release 4
  Goodheart-Wilcox Publishers (USA)
- Learning 3D Studio VIZ: A Tutorial Approach
  Goodheart-Wilcox Publishers (USA)
- Learning 3D Studio R4: A Tutorial Approach
  Goodheart-Wilcox Publishers (USA)

## CADCIM Technologies Textbooks Translated in Other Languages

### 3ds Max Textbook
- 3ds Max 2008: A Comprehensive Guide (Serbian Edition)
  Mikro Knjiga Publishing Company, Serbia

### SolidWorks Textbooks
- SolidWorks 2008 for Designers (Serbian Edition)
  Mikro Knjiga Publishing Company, Serbia
- SolidWorks 2006 for Designers (Russian Edition)
  Piter Publishing Press, Russia
- SolidWorks 2006 for Designers (Serbian Edition)
  Mikro Knjiga Publishing Company, Serbia
- SolidWorks 2006 for Designers (Japanese Edition)
  Mikio Obi, Japan

### NX Textbooks
- NX 6 for Designers (Korean Edition)
  Onsolutions, South Korea
- NX 5 for Designers (Korean Edition)
  Onsolutions, South Korea

## Pro/ENGINEER Textbooks
* Pro/ENGINEER Wildfire 4.0 for Designers (Korean Edition)
  HongReung Science Publishing Company, South Korea
* Pro/ENGINEER Wildfire 3.0 for Designers (Korean Edition)
  HongReung Science Publishing Company, South Korea

## AutoCAD Textbooks
* AutoCAD 2006 (Russian Edition)
  Piter Publishing Press, Russia
* AutoCAD 2005 (Russian Edition)
  Piter Publishing Press, Russia
* AutoCAD 2000 Fondamenti (Italian Edition)
* AutoCAD 2000 Tecniche Avanzate (Italian Edition)
* AutoCAD 2000 (Chinese Edition)

## Coming Soon from CADCIM Technologies
* AutoCAD 2014: A Problem-Solving Approach
* Autodesk Inventor 2014 for Designers
* AutoCAD Plant 3D 2014 for Designers
* Creo Direct 2.0 and Beyond for Designers
* Autodesk Simulation Mechanical 2014 for Designers
* NX Nastran 8.5 for Designers
* Pixologic ZBrush 4R5: A Comprehensive Guide
* Autodesk 3ds Max Design 2014: A Tutorial Approach
* Autodesk 3ds Max 2014: A Comprehensive Guide
* Autodesk Softimage 2014: A Tutorial Approach
* Exploring Autodesk Revit Architecture 2014 for Architects and Designers
* Autodesk Maya 2014: A Comprehensive Guide
* Exploring Autodesk Naviswork 2014
* Exploring AutoCAD Civil 3D 2014

## Online Training Program Offered by CADCIM Technologies
CADCIM Technologies provides effective and affordable virtual online training on various software packages including computer programming languages, Computer Aided Design and Manufacturing (CAD/CAM), animation, architecture, and GIS. The training will be delivered 'live' via Internet at any time, any place, and at any pace to individuals as well as the students of colleges, universities, and CAD/CAM training centers. For more information, please visit the following link: *http://www.cadcim.com*

www.ingramcontent.com/pod-product-compliance
Lightning Source LLC
LaVergne TN
LVHW062300060326
832902LV00013B/1986